EARTH SCIENCES IN THE 21ST CENTURY

YANGTZE RIVER

GEOGRAPHY, POLLUTION AND ENVIRONMENTAL IMPLICATIONS

EARTH SCIENCES IN THE 21ST CENTURY

Additional books in this series can be found on Nova's website under the Series tab.

Additional e-books in this series can be found on Nova's website under the e-book tab.

ENVIRONMENTAL SCIENCE, ENGINEERING AND TECHNOLOGY

Additional books in this series can be found on Nova's website under the Series tab.

Additional e-books in this series can be found on Nova's website under the e-book tab.

EARTH SCIENCES IN THE 21ST CENTURY

YANGTZE RIVER

GEOGRAPHY, POLLUTION AND ENVIRONMENTAL IMPLICATIONS

TRACY B. MALONEY
AND
BOYCE R. HUTCHINS
EDITORS

NOVINKA

New York

For permission to use material from this book please contact us:
Telephone 631-231-7269; Fax 631-231-8175
Web Site: http://www.novapublishers.com

Library of Congress Cataloging-in-Publication Data

ISBN: 978-1-62618-285-1

Published by Nova Science Publishers, Inc. † New York

CONTENTS

Preface **vii**

Chapter 1 Chlorinated Organic Contaminants in Surface Sediments of Yangtze River Estuary and Adjacent East China Sea **1**
Lili Niu, Huayun Yang and Weiping Liu

Chapter 2 Environmental and Land-use Changes in the Tibetan Plateau Section of the Upper Yangtze River Basin during the Last 50 Years **35**
Stuart A. Harris

Chapter 3 The Yangtze River: Hydro Development, Changing Geography, Cultural and Environmental Implications **65**
Kepa Morgan and Amaria Francis Waretini

Chapter 4 Innovative Solutions for Yangtze River's Water Crisis **89**
Shu-Qing Yang

Chapter 5 The Research of Environmental Flows Methodology in Rivers of China **109**
X. Ban

Chapter 6 Urban Development and its Impacts on Energy and Resource Consumptions in the Yangtze River Delta: Trends and Future Prospects **121**
Keishiro Hara, Michinori Uwasu, Helmut Yabar and Haiyan Zhang

Index **129**

PREFACE

The Yangtze River is Asia's longest river and the third longest river in the world. This book explores the Yangtze River's geography, pollution, and environmental implications. Topics discussed include chlorinated organic contaminants in surface sediments of the Yangtze River estuary and adjacent East China Sea; environmental and land-use changes in the Tibetan Plateau section of the Upper Yangtze River Basin during the last fifty years; hydro-development, the environmental and cultural sustainability of the Yangtze River; innovative solutions for the Yangtze River's water crisis; environmental flows research methodology in the rivers of China; and urban development and its impacts on energy and resource consumptions in the Yangtze River Delta.

Chapter 1 - The Yangtze River Estuary (YRE) and the adjacent East China Sea (ECS) are industrialized and urbanized regions in eastern China. The rampant development in the urban areas around has led to severe contamination here. In China, organochlorine pesticides (OCPs) and polychlorinated biphenyls (PCBs) were extensively used in agricultural activities and for industrial applications between the 1950s and 1980s and between the 1960s and 1970s, respectively. Although the use of these chlorinated organic contaminants has been restricted or banned for several years, they are still frequently detected. These contaminants are of wide concern due to their persistence in the environment, transportation between phases, bioaccumulation in the food web and threat to human and ecosystem health. Thus, they are classified as persistent organic pollutants (POPs). In addition, as a secondary source of anthropogenic contaminants, estuarine sediments are also the primary OCP and PCB reservoirs in the YRE. The sediments provide valuable information regarding the aquatic environment pollution status. In this chapter, a comprehensive status of chlorinated organic pollution in the YRE and the adjacent ECS surface sediments is presented. In this summary, the authors focus on the pollution, distribution, potential source and ecotoxicological assessment of OCPs (DDT, hexachlorocyclohexanes and others) and PCBs. In addition, this chapter highlights the chiral signatures of the OCPs, which can be used to trace their pathways. This method provides better description of these contaminants in the YRE and adjacent ECS sediments.

Chapter 2 - The upper Yangtze River basin can be divided into a western section including the Qinghai-Tibetan Plateau and the eastern slopes down to Yibin City, and the eastern region above the Three Gorges Dam. This chapter discusses the environmental and land-use changes that have taken place during the last 50 years in the western portion. The Yangtze River flows from the terminus of a glacier at 6,500 m on the Tanggula Range, crossing the northwest part of the Qinghai-Tibetan Plateau at an elevation between 4700 m and 3800 m. This is a cold, arid area with a semi-desert vegetation, currently undergoing desertification due to climate change and recent land-use practices. The river then descends slowly across cold high-altitude *Kobresia* meadows before plunging down to about 300 m elevation through spectacular gorges with a mixed coniferous forest vegetation. Exploitation of forests, mining and construction of dams has and is modifying these gorges, though several substantial areas are now protected. Reforestation of mountain sides is being conducted to try to revegetate the steep slopes and reduce erosion. Mining is now restricted to specific areas, and there are resettlement plans for the people displaced by the proposed construction of a sequence of cascading dams planned to provide much needed electricity to meet the demands of an increasingly industrialized nation.

Chapter 3 - The Yangtze River system is Asia's longest river and has its glacial source in Qinhai province on the Tibetan Plateau. The Yangtze flows eastward from its source 6,418 kilometres across the People's Republic of China (China) before discharging into the East China Sea. The Yangtze catchment is one-fifth of the total land area of China, making the Yangtze River one of the largest in the world in terms of discharge volume.

The Yangtze and over 700 tributaries continue to be the primary transportation link for one third of China's population (excluding the Grand Canal connectivity), and the main link for international shipping to the interior provinces of China. The rivers of primary importance to millions of people (more than thirty cities are located along the river) living on adjacent lands including the flood plains downstream of the Three Gorges Project. The river is administered by the Yangtze Forum which brings together 13 riparian provincial governments managing the river from its headwaters to the East China Sea. Many cities extract and treat its waters for potable use, and improved riparian management encompassing reconstituting and reconnecting the myriad of disconnected lakes, watercourses and wetlands has protected the drinking water sources for several cities and also enhanced ecosystem habitat for many threatened species.

Despite improvements in the decision making processes, the Yangtze River remains the focus of on-going tensions regarding the sustainability and the appropriateness of the Three Gorges Project development completed in 2011. The Yangtze River system has been irreversibly altered from the "Golden Waterway" to a strategic asset now producing one-sixth of China's total electrical generating capacity supporting more than 20% of China's GDP. The Three Gorges Project is a critical part of the plan China is implementing to increase its contribution to world trade and the standard of living for its people. River flow has been extensively modified with resulting implications in terms of reduced flood risks and sediment loads downstream, while the significantly raised water levels in the dam reservoir have forced the relocation of more than 1.2 million people and impacted upon geological stability.

The combination of historic relevance, impacted anthropological sites, ecosystem modification, altered river flows, industrial and urban uses, power generation, transportation reliability, geological instability, siltation issues, and economic base, under the Western World's spotlight, are some of the inter-related attributes that make the Yangtze River a complex management challenge: A contemporary interpretation perhaps of its metaphor; the Golden Waterway.

The Mauri Model Decision Making Framework offers a new approach to the analysis of water resource management as it provides a transparent and inclusive approach to considering the environmental, economic, social and cultural aspects of decisions that have been taken regarding the Yangtze River system. The Mauri Model Decision Making Framework is capable of understanding multiple-worldviews and adopts mauri (intrinsic value, well-being or Qi) in the place of monetary equivalents as the base metric allowing an absolute sustainability assessment.

There are interesting parallels between the Yangtze River and the Waikato River in New Zealand, as while the actual physical scales are demonstrably different, the contextual relevance of the rivers resonates, as does the strategic importance of these rivers to their respective governments. The changing perceptions of both rivers from different worldviews, offers new ways of understanding sustainability, and better understanding the cultural and environmental implications of the development taking place on the Yangtze River.

A sustainability assessment for the Yangtze River focusing on the contemporary changes associated with the period of the Three Gorges Project is presented and future opportunities are identified. The application of the Mauri Model Decision Making Framework to the Yangtze River reveals the parallels between the concepts of mauri and Qi, and the potential opportunities for the application of the framework in other Asian and Pacific contexts.

Chapter 4 - Yangtze River is one of the largest rivers in the world, it has disasters like floods, droughts, sedimentation and water pollution etc. After the operation of Three-Gorge-Dam (TGD), the largest dam in the basin, the water shortage problem becomes very severe and frequent, especially in dry seasons. This chapter's analysis shows that the application of SPP scheme in Dongting and Poyang Lakes in the river basin can successfully eliminate these disasters. The so called SPP stands for "separation, prevention and protection", the chapter presents how to apply the SPP to these lakes. The feasibility is discussed in terms of technology, economical affordability and environmental sustainability. The chapter also compares the SPP scheme with the scheme of sluice gates/dam at the lake's outlets which the local governments have been put forward, the results show that the payback periods of two schemes are almost same, but the SPP scheme is much better than the gate/dam scheme in every aspect. It is found that the disasters in the river basin can be effectively eliminated if the proposed SPP is applied in the basin's lakes.

Chapter 5 - Environmental Flows (EFs) (Troldborg et al.) refers to the basic flow rate and process retained in the river for maintaining the river ecosystem. At present, there are many methods to determine the EFs, so it has become a frontier and hotspot issue now to choose the right EFs suitable for the characteristics of rivers in China. There are many rivers in China, with great difference in natural environment and ecosystem. Especially due to different levels of human development and utilization, it is difficult to determine the EFs with uniform standards, and different measures should be taken based on the difference in rivers and ecological protection goals. Based on the idea of river zoning and classification, by using research achievements at home and abroad, and based on different ecosystems of rivers in China, this paper classifies the EFs problems into three spatial levels: whole country, river basin and water function. Based on the characteristics of rivers in different regions and different reaches and in combination with ecological and environmental problems, different methods are used to calculate and determine

EFs. This paper comprehensively summarizes the research of EFs in China, and recommends using different EFs calculation methods based on the characteristics of rivers in China, aiming to provide a scientific guidance for the research in China.

Chapter 6 - Due to its rapid industrialization and urbanization, China is facing the challenge of sharply growing energy and resource consumption, while it simultaneously faces various socio-economic challenges such as growing income gaps between urban and rural areas. These challenges are posing threats to regional sustainability. In this paper the authors look into the Yangtze River Delta region which consists of Shanghai city and the neighboring provinces of Jiangsu and Zhejiang — the fastest economically developing region of the country. The authors first review recent trends in urbanization and industrialization of the region. They then demonstrate that the rapid urbanization and associated industrial activities have had tremendous impacts on energy and resources consumption, thus having an impact on regional sustainability. The authors finally discuss that technological innovations in secondary industries and pursuing industrial symbiosis in the context of circular economy would be of vital importance especially for the industrial sectors of the region to pursue sustainable development.

In: Yangtze River ISBN: 978-1-62618-285-1
Editors: T. B. Maloney, B. R. Hutchins

Chapter 1

CHLORINATED ORGANIC CONTAMINANTS IN SURFACE SEDIMENTS OF YANGTZE RIVER ESTUARY AND ADJACENT EAST CHINA SEA

Lili Niu, Huayun Yang and Weiping Liu*

International Joint Research Centre for Persistent Toxic Substances, MOE Key Laboratory of Environmental Remediation and Ecosystem Health, College of Environmental and Resource Sciences, Zhejiang University, Hangzhou, China

ABSTRACT

The Yangtze River Estuary (YRE) and the adjacent East China Sea (ECS) are industrialized and urbanized regions in eastern China. The rampant development in the urban areas around has led to severe contamination here. In China, organochlorine pesticides (OCPs) and polychlorinated biphenyls (PCBs) were extensively used in agricultural activities and for industrial applications between the 1950s and 1980s and between the 1960s and 1970s, respectively. Although the use of these chlorinated organic contaminants has been restricted or banned for several years, they are still frequently detected. These contaminants are of wide concern due to their persistence in the environment, transportation between phases, bioaccumulation in the food web and threat to human

---

* Corresponding author: Weiping Liu, Tel./Fax: +86-571-88982341. E-mail addresses: wliu@zju.edu.cn.

and ecosystem health. Thus, they are classified as persistent organic pollutants (POPs). In addition, as a secondary source of anthropogenic contaminants, estuarine sediments are also the primary OCP and PCB reservoirs in the YRE. The sediments provide valuable information regarding the aquatic environment pollution status. In this chapter, a comprehensive status of chlorinated organic pollution in the YRE and the adjacent ECS surface sediments is presented. In this summary, we focus on the pollution, distribution, potential source and ecotoxicological assessment of OCPs (DDT, hexachlorocyclohexanes and others) and PCBs. In addition, this chapter highlights the chiral signatures of the OCPs, which can be used to trace their pathways. This method provides better description of these contaminants in the YRE and adjacent ECS sediments.

Keywords: Organochlorine pesticides, polychlorinated biphenyls, sediment, Yangtze River Estuary, East China Sea, Chirality

INTRODUCTION

The Yangtze River Estuary (YRE) and the adjacent East China Sea (ECS) are major commercial arteries. Combined, these areas make up one of the most industrialized and urbanized regions in China. Cities and provinces like Shanghai and Zhejiang exist along the YRE and ECS coasts. These cities have large populations that adversely affect the YRE and ECS environmental conditions for long periods of time. Historically, pesticide applications in Zhejiang were heavy as a result of agricultural development. From 1985 to 1991, the pesticide consumption rate in Zhejiang agriculture zone reached approximately 9.96 kg/hm^2, which was 2.7-fold greater than the national average (2.71 kg/hm^2) (Wang et al. 2005). Moreover, the total application of organochlorine pesticides in the Zhejiang Province accounted for one-tenth of the total application in China between 1952 and 1984 (Wu et al. 2007). These pesticides may enter into and adversely impact the nearby YRE and ECS ecosystems through leach and runoff. Taizhou is one of the largest e-waste recycling locations in the Zhejiang Province, China. Previously, large quantities of PCBs contained in dismantled e-wastes caused severe contamination on the environment quality in Taizhou and in the nearby YRE and ECS coastal areas (Chen et al. 2010; Han et al. 2010; Shen et al. 2008). The YRE is a sink for fine-grained sediments and associated particle-reactive pollutants. In addition, these sediments are an important source of hazardous contaminants for coastal marine environments (Feng et al.1998). Although the contaminant concentrations are within the specified water pollution criteria, these polluted sediments in estuarine and coastal areas may have adverse biological affect (Daskalakis and Oconnor 1995). Hence, the contamination status of the YRE and the adjacent ECS should be determined to manage and control environmental quality.

Chlorinated organic contaminants, such as polychlorinated biphenyls (PCBs) and organochlorine pesticides (including hexachlorocyclohexane (HCHs), chlordane and dichlorodi-phenyltrichloroethanes (DDTs)), are persistent organic pollutants (POPs). These pollutants receive global attention because of their environmental persistence, biomagnification and adverse ecosystem and human health effects (Ren et al. 2007; Zhang et al. 2007). Furthermore, these contaminants may enter the aquatic environment through various pathways, including atmosphere transportation and deposition and industrial wastewater discharge and surface runoff (Totten et al. 2006; Zhou et al. 2001). Due to their low water solubility and their high octanol-water partition coefficients (K_{ow}), theses contaminants can strongly absorb to suspended particulate matter, and accumulate in sediments. Thus, the sediment is an important reservoir or “sink” for these contaminants, which can lead to significant long-term problems (Bettinetti et al. 2003). Once disturbed, the sediment can be resuspended and serve as a secondary contamination source for other sites (Yang et al. 2005; Yang et al. 2009). Subsequently, the contaminants in sediments will enter into sediment-dwelling organisms and bioaccumulate in higher trophic levels through the food web (Liu et al. 2003). Therefore, information regarding the contaminant levels in sediments is useful for describing the status and accumulation of organic pollutants. In addition, this information can be used to provide a comprehensive risk assessment of the aquatic environment.

From the 1950s to the 1980s, DDTs, a family of DDT and its degradation products (such as DDD and DDE), and HCHs, (including α-, β-, γ-, and δ-HCH) were largely produced and used as organochlorine pesticides in China (Tao et al. 2008). Total Chinese production of HCHs and DDTs were estimated to exceed 4.9 million and 0.4 million tons, respectively, which accounted for 33% and 20% of their global production (Zhang et al. 2002). Although the use of OCPs was restricted or ceased in the early 1970s in many countries, high residue levels were still detected in various environmental matrixes (Aigner et al. 1998; Lin et al. 2012). In addition, some tropical and subtropical countries still use DDT, HCH, and lindane for agricultural and medicinal purposes (de Beito et al. 2002). These contaminants may enter other areas through atmosphere transportation, and thus still threaten the ecosystem and human health.

Approximately 25% of the currently used pesticides are chiral, including *α*-HCH, *o,p'*-DDT, *o,p'*-DDD, *trans*- (TC) and *cis*-chlordane (CC). However, most of these pesticides are manufactured and sold as racemic mixtures. Chiral pesticides have one or more pairs of enantiomers with the same physicochemical properties but with different biological effects. Microorganisms in sediments can preferentially degraded one enantiomer (Genualdi et al. 2009).

Thus, the enantiomeric ratios will deviate from 1:1. This preferential degradation can result in non-racemic signatures, which can be used to discriminate volatilized residues from atmospheric transport (Buser and Muller 1995). Therefore, the study of chiral OCPs at an enantiomeric level can provide valuable information regarding the release, distribution and deposition of contaminants at a site. In addition, an improved risk assessment for aquatic environments can be presented at an enantiomeric level.

PCBs were first manufactured commercially in 1929 and were widely used in various industrial applications, such as in the dielectric fluid in capacitors and transformers, in waxes, inks and pesticides additives (Miller 1982). The physicochemical stability, low flammability and electrical insulating properties of PCBs make them ideal for industrial materials (PedersenBjergaard et al. 1996).

Prior to inhibiting the production and use of PCBs in the early 1970s, more than 1.3×10^6 tons of PCBs were produced worldwide (Breivik et al. 2002). In China, the total amount of technical PCBs produced from 1965 to 1974 was approximately 1000 tons, which accounted for 0.6% of the total global production (Yamashita et al. 2000). Electric and electronic wastes (e-waste) were considered the main sources of PCBs after their production and use being banned. China was the largest e-waste recipient in the world. Thus, the occurrence of PCBs was routinely detected in the air, soil and sediment throughout China (Ren et al. 2007; Yang et al. 2009; Zhang et al. 2008).

In this chapter, we compiled 10 different studies that detailed the occurrence and historical trends of chlorinated organic contaminants in the YRE and the adjacent ECS surface sediments. This compilation will provide a comprehensive description and risk assessment of the OCPs and PCBs in the YRE and the adjacent ECS sediments.

Table 1. Residue concentrations of DDTs in the YRE and the adjacent ECS surface sediments (ng/g)

		DD-Ts	p,p'-DDE	o,p'-DDT	o,p'-DDD	p,p'-DDD	p,p'-DDT	(DDE+DDD)/DDT	DDD/DDE	o,p'-DDT/p,p'-DDT	Ref.
YRE & AS[a] n=50	Mean	1.96	0.40	0.59	0.19	0.27	0.50	0.49	1.50	1.34	Yang 2011
	Median	1.94	0.42	0.50	0.20	0.24	0.37	0.48	1.23	0.94	
	Min	0.27	0.04	0.03	0.02	0.05	0.06	0.26	0.51	0.26	
	Max	4.54	1.17	1.83	0.40	1.10	2.14	0.76	3.44	9.49	
	SD	1.16	0.26	0.46	0.10	0.19	0.46	0.13	0.78	1.37	
	CV	0.59	0.65	0.79	0.53	0.70	0.92	0.26	0.52	1.02	
YE and CA[b] n=14	Mean	0.07									Liu et al. 2004
	Median	nd									
	Min	nd									
	Max	0.57									
	SD	0.17									
	CV	2.31									
YE and CA[c] n=14	Mean	0.07	0.03			0.03	0.02				Liu et al. 2003
	Median	nd	nd			nd	nd				
	Min	nd	nd			nd	nd				
	Max	0.56	0.35			0.34	0.21				
	SD	0.17	0.09			0.09	0.06				
	CV	2.31	3.74			2.86	3.74				

Table 1. (Continued)

		DD-Ts	p,p'-DDE	o,p'-DDT	o,p'-DDD	p,p'-DDD	p,p'-DDT	(DDE+DDD)/DDT	DDD/DDE	o,p'-DDT/p,p'-DDT	Ref.
YE[d] n=10	Mean	8.80	1.92	1.59		4.73	0.57	0.78	3.57	2.95	Liu et al. 2008
	Median	3.20	0.80	0.90		1.85	nd	0.72	2.32	1.37	
	Min	0.90	nd	nd		0.90	nd	0.58	1.40	0.00	
	Max	33.10	8.60	10.90		13.20	2.50	1.00	8.50	9.08	
	SD	10.14	2.65	3.31		4.73	0.89	0.17	2.58	4.20	
	CV	1.15	1.38	2.08		1.00	1.56	0.21	0.72	1.42	
YQB[e] n=6	Mean	7.25	0.60	3.82	0.76	0.55	1.51	0.40	2.14	2.82	Yang et al. 2010
	Median	5.60	0.54	3.10	0.82	0.37	1.02	0.37	2.18	2.30	
	Min	1.85	0.20	0.92	0.26	0.17	0.30	0.29	1.48	1.93	
	Max	16.54	1.01	9.03	1.24	1.65	3.61	0.61	2.86	5.37	
	SD	5.19	0.29	2.95	0.33	0.55	1.23	0.12	0.47	1.32	
	CV	0.72	0.47	0.77	0.43	0.99	0.82	0.29	0.22	0.47	
SMB[f] n=5	Mean	4.57	0.70	1.86	0.38	0.39	1.24	0.49	1.11	1.60	
	Median	4.47	0.70	1.48	0.39	0.30	1.03	0.50	1.13	1.22	
	Min	2.77	0.46	0.91	0.31	0.18	0.91	0.42	0.84	0.70	
	Max	6.93	0.98	3.64	0.44	0.75	2.10	0.56	1.35	3.03	
	SD	1.59	0.21	1.07	0.05	0.22	0.49	0.06	0.18	0.95	
	CV	0.35	0.30	0.58	0.13	0.57	0.40	0.12	0.17	0.59	

		DD-Ts	p,p'-DDE	o,p'-DDT	o,p'-DDD	p,p'-DDD	p,p'-DDT	(DDE+DDD)/DDT	DDD/DDE	o,p'-DDT/p,p'-DDT	Ref.
XSB[g] n=8	Mean	6.58	1.06	2.21	0.30	0.25	2.76	0.32	0.68	1.22	Yang et al. 2011
	Median	2.83	0.62	1.04	0.27	0.20	0.61	0.34	0.79	1.30	
	Min	0.61	0.14	0.24	0.01	0.06	0.16	0.18	0.22	0.33	
	Max	22.38	3.22	8.35	0.83	0.63	10.12	0.42	0.96	1.76	
	SD	7.49	1.05	2.74	0.24	0.19	3.58	0.09	0.27	0.51	
	CV	1.14	0.99	1.24	0.81	0.73	1.30	0.29	0.39	0.42	
ECS[h] n=8	Mean	3.10	1.00	0.18		1.05	0.98			0.20	Yang et al. 2005
	Median	3.46	1.16	nd		1.00	0.72			0.22	
	Min	nd	nd	nd		nd	nd			0.03	
	Max	6.07	2.05	0.44		2.21	2.92			0.41	
	SD	2.16	0.64	0.15		0.76	1.04			0.15	
	CV	0.70	0.64	0.83		0.72	1.07			0.75	

[a] YRE & AS is Yangtze River Estuary and adjacent Sea; [b, c] YE & CA is Yangtze Estuary and nearby Coastal Areas; [d] YE is Yangtze Estuary; [e] YQB is Yueqing Bay; [f] SMB is Sanmen Bay; [g] XSB is Xiangshan Bay; [h] ECS is East China Sea.

Pollution of DDTs in the YRE and the Adjacent ECS Surface Sediments

Residue Levels and Distribution Characteristics

The residual isomer concentrations of DDT in the YRE and the adjacent ECS surface sediments from previous studies were summarized in Table 1. The total DDT concentrations from these studies were up to 33.10 ng/g (based on dry weight), with average levels of between 0.07 and 8.80 ng/g. In studies conducted by Yang et al. (Yang 2011; Yang et al. 2010, 2011), the DDT residues were detected in all of the sediment samples, which indicated that DDT pollution was widespread in the selected sites. The significantly variable DDT concentrations detected in the YRE and the adjacent ECS primarily originated from different locations. The highest ∑DDT concentrations occurred in the Yangtze Estuary (YE) sediment near an important sewage discharge mouths. In addition, Yang (2011) found the highest DDT concentrations in the surface sediment of the Yangtze River Estuary and the adjacent sea (YRE & AS) presumably in the river-sea boundary zone known as the "marginal filter". This zone is a narrow belt where water from small and large rivers mixes with sea water, resulting in geochemical and hydrodynamic processes, such as sorption, coagulation-flocculation and sedimentation. It was reported that more than 90% of the suspended matter and 40% of the dissolved elements in the river flux potentially deposit in this zone (Nemirovskaya 2009). Therefore, abundant contaminants like DDTs accumulated here. The second location with higher DDT concentrations was near the inner mouth of the Yangtze and the Qiantang River. This location received contamination from heavy industrial and commercial activities and the pollutants were hardly to be diluted or cleaned. Furthermore, except the geographical characteristics of the regions, which mainly contributed to the spatial distribution of the contaminants in the sediments, the hydrodynamic factors and the aquatic environmental properties might also play an important role in the accumulation of DDTs in sediments.

Composition and Potential Source

The dominant isomer detected in the YRE & AS, Yueqing Bay (YQB), Sanmen Bay (SMB) and Xiangshan Bay (XSB) was *o,p'*-DDT, followed by *p,p'*-DDT (Figure 1). However, in the Yangtze Estuary (YE) and East China Sea (ECS) studies, the most common isomer was *p,p'*-DDD and *p,p'*-DDE, respectively. This result was potentially caused by the difference DDT sources in these regions. Therefore, we calculated the relevant concentrations of the parent DDT, its metabolites and its different isomers to trace the potential sources of DDTs in the YRE and the adjacent ECS sediments. The *o,p'*-DDT/*p,p'*-DDT ratio in technical DDT produced and used before was determined to fall between 0.2 and 0.3 (Kalantzi et al. 2001). However, the *o,p'*-DDT/*p,p'*-DDT ratio in dicofol, an insecticide substituted for the banned DDT, is between 1.3 and 9.3 or greater (Qiu et al. 2005). As shown in Table 1, the average *o,p'*-DDT/*p,p'*-DDT ratios in the surface sediments were much higher than 0.3 in the studies conducted by Yang et al. (Yang 2011; Yang et al. 2010, 2011) and Liu et al. (2008), indicating sources from dicofol occurred. The surface sediment (0 cm) of the ECS had a mean ratio of 0.2 and a maximum ratio of 0.41. This result suggested a historical residue from technical DDT presented in the ECS sediments since *o,p'*-DDT is unstable in the environment.

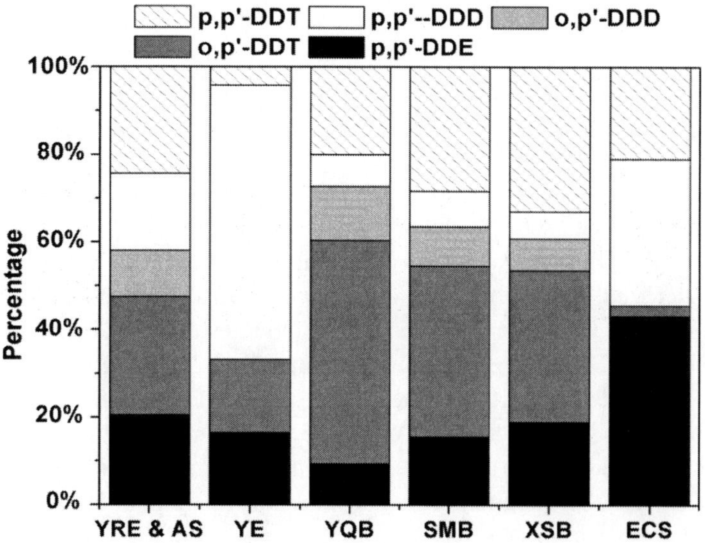

Figure 1. DDT composition in the surface sediments of Yangtze River Estuary and adjacent Sea (YRE & AS) (Yang 2011), Yangtze Estuary (YE) (Liu et al. 2008), Yueqing Bay (YQB) (Yang et al. 2010), Sanmen Bay (SMB) (Yang et al. 2010), Xiangshan Bay (XSB) (Yang et al. 2011) and East China Sea (ECS) (Yang et al. 2005). In the studies in YE and ECS, the determination of *o,p'*-DDD was absent.

DDE and DDD are two types of DDT metabolites that occur under aerobic and anaerobic conditions, respectively (Bopp et al. 1982). The DDD/DDE ratio can be used to determine the environmental conditions during DDT decomposition. For example, a ratio > 1 indicates that the DDT was degraded under anaerobic condition and a ratio < 1 indicates that the DDT was degraded under aerobic condition (Hitch and Day 1992). The average DDD/DDE ratios in the YRE & AS, YE, YQB and SMB were all > 1 (Table 1). This finding suggested that anaerobic condition occurred during DDT degradation, which was potentially caused by abundant organic matter from years of fish farming. Furthermore, the (DDE+DDD)/DDT ratio can indicate if DDT emission occurred recently or historically (Doong et al. 2002). A (DDE+DDD)/DDT ratio > 0.5 means that the sediment contaminants were weathered over a long period of time. Most ratios < 0.5 were observed in the YRE & AS, YQB, SMB and XSB sediments, which suggested that recent DDT inputs occurred in these areas. In the YE sediments, the average ratio was slightly higher than 0.5, which might be due to the past usage.

Ecotoxicological Risk

The DDTs in polluted sediments may have adverse effects on sediment-dwelling organisms, which may subsequently enrich in higher trophic level animals through the food web. It will pose a great threat to ecosystem and human health when exposed to DDTs. Thus, the potential risks of sediment contamination with DDTs should be assessed. Herein, sediment quality guidelines (SQG) were suggested and comparisons were performed between current data with two criteria, i.e., effects range-low (ERL) and effects range-median (ERM). Therefore the potential ecotoxicities of OCPs in the YRE and the adjacent ECS surface sediments can be predicted (Long et al. 1995). The ERL refers to the lower 10% of the effects data of the chemicals, while ERM refers to 50% of the effects data. When the contaminant levels are below the ERL, minimal adverse effects occur. In addition, probable-effects frequently occur when the concentration of pollutants is equivalent to or above the ERM. The quality criteria of *p,p*'-DDT, *p,p'*-DDD, *p,p'*-DDE, and total DDT, and the frequency of their concentrations in each criterion range are shown in Figure 2. All of the DDT concentrations in the Yangtze Estuary and nearby Coastal Areas (YE and CA) sediments were below the ERL, indicating that adverse influence rarely happened. The *p,p'*-DDT levels in most YRE and the adjacent ECS sediments were in the intermediate range. Only samples from the XSB were greatly threatened by *p,p'*-DDT. For *p,p'*-DDD and *p,p'*-DDE, all of the samples were in the safe range (< ERL), with the exception of those from the XSB, ECS and YE. This finding revealed that risks from *p,p'*-DDD and *p,p'*-DDE occurred occasionally in the XSB, EC and YE sediments. However, the other sites were relatively safe. Most of samples had intermediate sediment toxicity for DDTs since more than 50% of the sediments had total DDT concentrations between the ERL and ERM, except for the toxicity rating for the YE and CA.

Table 2. Residue concentrations of HCHs in the YRE and the adjacent ECS surface sediments (ng/g)

		HCHs	α-HCH	β-HCH	γ-HCH	δ-HCH	α-HCH/γ-HCH	Ref.
YRE & AS[a] n=50	Mean	0.76	0.15	0.16	0.02	nd	6.48	Yang 2011
	Median	0.35	0.14	0.16	0.02	0.08	6.07	
	Min	0.10	0.05	0.04	0.01	nd	3.46	
	Max	21.00	0.40	0.36	0.08	0.08	11.66	
	SD	2.92	0.07	0.07	0.01	0.02	2.02	
	CV	3.82	0.44	0.44	0.58	0.37	0.31	
YE and CA[b] n=14	Mean	7.18						Liu et al. 2004
	Median	3.19						
	Min	nd						
	Max	32.63						
	SD	10.79						
	CV	1.50						
YE and CA[c] n=14	Mean	7.19	nd	0.07	6.76	0.39	0.01	Liu et al. 2003
	Median	3.19	nd	nd	3.04	nd	0.00	
	Min	nd	nd	nd	nd	nd	0.00	
	Max	32.63	0.10	0.72	30.40	5.42	0.05	
	SD	10.79	0.03	0.19	9.88	1.45	0.02	
	CV	1.50	2.60	2.63	1.46	3.68	2.48	
YE[d] n=10	Mean	6.04	3.28	0.33	0.73	1.72	4.74	Liu et al. 2008
	Median	5.20	1.50	0.30	0.60	1.30	2.56	
	Min	0.50	0.50	nd	nd	nd	0.80	
	Max	17.50	14.30	1.20	1.70	5.50	18.00	
	SD	4.66	4.35	0.38	0.59	1.90	5.88	
	CV	0.77	1.33	1.15	0.81	1.11	1.24	

Table 2. (Continued)

		HCHs	α-HCH	β-HCH	γ-HCH	δ-HCH	α-HCH/γ-HCH	Ref.
YQB[e] n=6	Mean	0.91	0.40	0.42	0.06	0.03	7.28	Yang et al. 2010
	Median	0.86	0.38	0.42	0.05	0.02	7.10	
	Min	0.34	0.15	0.15	0.03	0.01	5.00	
	Max	1.55	0.72	0.64	0.11	0.09	11.70	
	SD	0.41	0.21	0.20	0.03	0.03	2.47	
	CV	0.45	0.52	0.49	0.58	0.88	0.34	
SMB[f] n=5	Mean	0.55	0.24	0.25	0.04	0.02	7.18	
	Median	0.46	0.20	0.20	0.03	0.02	6.70	
	Min	0.37	0.13	0.18	0.02	0.01	5.90	
	Max	0.97	0.41	0.46	0.07	0.03	9.50	
	SD	0.24	0.11	0.12	0.02	0.01	1.39	
	CV	0.44	0.45	0.48	0.58	0.38	0.19	
XSB[g] n=8	Mean	0.39	0.14	0.21	0.02	0.02	8.74	Yang et al. 2011
	Median	0.35	0.11	0.17	0.02	0.00	6.00	
	Min	0.14	0.00	0.14	0.00	0.00	2.33	
	Max	0.67	0.28	0.36	0.05	0.05	26.00	
	SD	0.19	0.10	0.09	0.02	0.02	8.85	
	CV	0.50	0.73	0.44	0.89	1.43	1.01	
ECS[h] n=8	Mean	0.86	0.22	0.49	0.06	nd	4.31	Yang et al. 2005
	Median	0.86	nd	0.51	nd	nd	3.50	
	Min	0.20	nd	nd	nd	nd	0.83	
	Max	1.45	0.40	0.94	0.13	0.15	11.00	
	SD	0.38	0.14	0.34	0.05	0.03	3.18	
	CV	0.44	0.62	0.69	0.73	0.45	0.74	

[a] YRE & AS is Yangtze River Estuary and adjacent Sea; [b, c] YE & CA is Yangtze Estuary and nearby Coastal Areas; [d] YE is Yangtze Estuary; [e] YQB is Yueqing Bay; [f] SMB is Sanmen Bay; [g] XSB is Xiangshan Bay; [h] ECS is East China Sea.

Generally, most samples were within the safe or in the intermediating range. Thus, the risks of DDTs shown in the YRE and the adjacent ECS surface sediments were low or moderate. However, DDT contamination should be paid great concern because DDTs degrade slowly and may biomagnify in organisms through the food web.

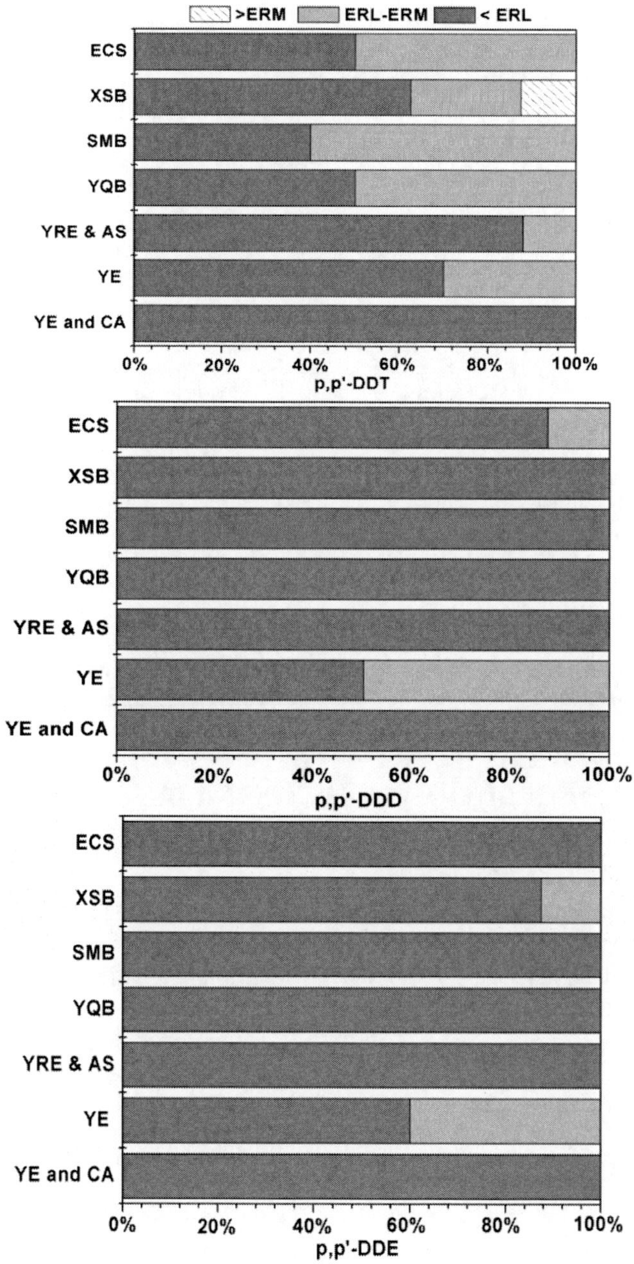

Figure 2. (Continued).

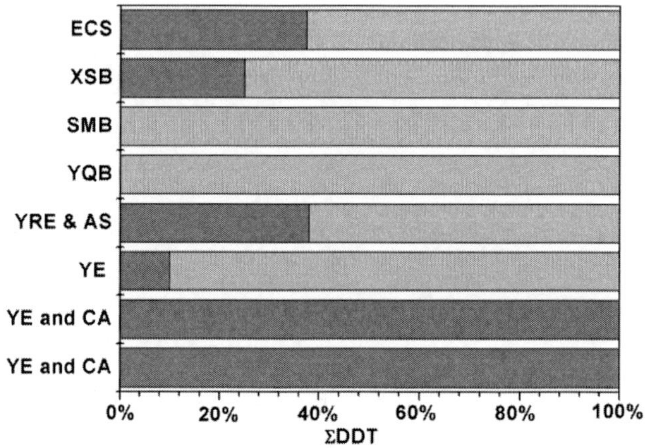

Figure 2. Frequency of reported concentrations in the three criterion ranges for DDTs in the surface sediments of Yangtze Estuary and nearby Coastal Areas (YE and CA) (Liu et al. 2003, 2004), Yangtze Estuary (YE) (Liu et al. 2008), Yangtze River Estuary and adjacent Sea (YRE & AS) (Liu et al. 2008), Yueqing Bay (YQB) (Yang et al. 2010), Sanmen Bay (SMB) (Yang et al. 2010), Xiangshan Bay (XSB) (Yang et al. 2011) and East China Sea (ECS) (Yang et al. 2005). The values of effect range-low (ERL) for *p,p'*-DDT, *p,p'*-DDD, *p,p'*-DDE and ∑DDT are 1, 2, 2.2 and 1.58 ng/g, respectively. The values of effect range-median (ERM) for *p,p'*-DDT, *p,p'*-DDD, *p,p'*-DDE and ∑DDT are 7, 20, 27 and 46.1 ng/g, respectively (Long et al. 1990, 1995).

Pollution of HCHs in the YRE and the Adjacent ECS Surface Sediments

Residue Levels and Distribution Characteristics

The α-HCH, β-HCH, γ-HCH, δ-HCH and total HCH residue levels in the YRE and the adjacent ECS surface sediments were determined and are listed in Table 2. The detection rate of HCHs was 100% in these regions, except for at the YE and CA, which indicated that the HCH pollution was ubiquitous in the YRE and the adjacent ECS sediments. These studied areas had a ∑HCHs burden of between nd and 32.63 ng/g, with mean values between 0.39 and 7.19 ng/g (dry weight). The HCH contents in the sediments varied widely among the different sampling locations. The highest total HCH concentration was found in the YE and CA sites. Several of these sites were influenced by nearby sewage discharge points. Similar to the spatial distribution of DDTs, higher levels of HCHs were also found in the marginal filter and the inner river outlets (Yang 2011). In addition, sampling time may influence the residue levels of HCHs in sediments. The higher HCH concentrations occurred in the earlier studies in the YE (sampled in 2001) and YRE and CA (sampled in 2002). After years of degradation in or leaching from sediments, the HCH concentrations decreased. The total HCH contents were all < 1 ng/g in the YRE & AS, YQB, SMB, XSB and ECS. Thus, contamination greatly decreased following the prohibition of HCHs in the 1970s. In the cases of the YRE & AS, YQB, SMB, XSB and ECS, the HCH concentrations were much lower than the DDT concentration, although the production of DDTs was only about one tenth that of HCHs between 1950s and 1983 in China (Zhang et al. 2002). It was likely due to the higher water solubility, biodegradability and vapor pressure, but lower particle affinity and lipophilicity of the HCHs relative to DDTs (Yang et al. 2005).

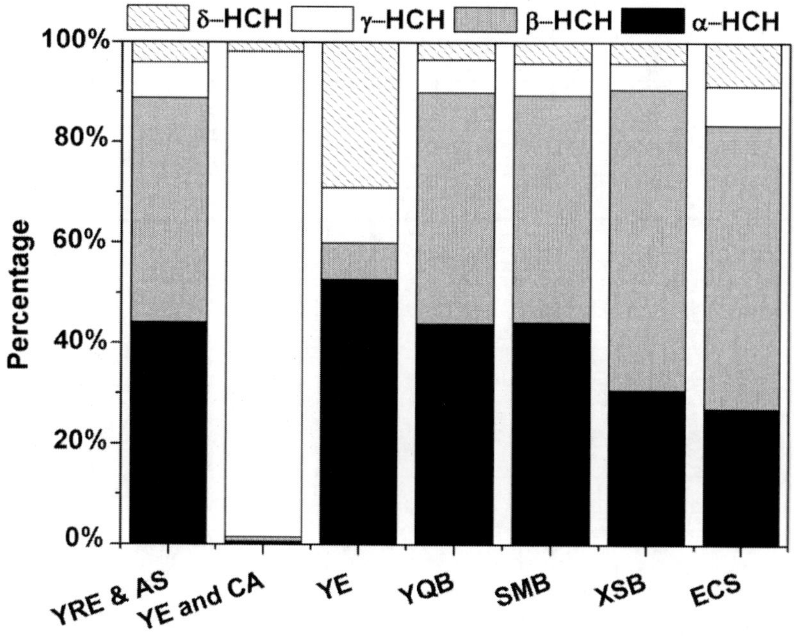

Figure 3. HCH composition in the surface sediments of Yangtze River Estuary and adjacent Sea (YRE & AS) (Yang 2011), Yangtze Estuary and nearby Coastal Areas (YE and CA) (Liu et al. 2003), Yangtze Estuary (YE) (Liu et al. 2008), Yueqing Bay (YQB) (Yang et al. 2010), Sanmen Bay (SMB) (Yang et al. 2010), Xiangshan Bay (XSB) (Yang et al. 2011) and East China Sea (ECS) (Yang et al. 2005).

Table 3. Residue concentrations of other OCPs in the YRE and the adjacent ECS surface sediments (ng/g)

		Chlor-dane	CC	TC	TC/CC	Ald-rin	Diel-drin	Hepta-chlo	Heptac-hlor-Epoxide	Endo-sulfan I	Endosu-lfan II	Endri-n/Ketone	Ref.
YRE & AS[a] n=50	mean	0.12	0.04	0.08	1.98								Yang 2011
	median	0.13	0.04	0.08	1.83								
	min	nd	nd	nd	0.00								
	max	0.22	0.09	0.18	5.00								
	SD	0.06	0.01	0.05	1.25								
	CV	0.44	0.31	0.62	0.63								
YE and CA[b] n=14	mean					0.77	0.14	0.13	0.06	0.15	0.03	nd	Liu et al. 2004
	median					0.65	0.16	nd	nd	nd	nd	nd	
	min					0.22	nd	nd	nd	nd	nd	nd	
	max					2.48	0.27	0.93	0.31	1.23	0.16	0.01	
	SD					0.63	0.08	0.32	0.12	0.34	0.06	0.00	
	CV					0.82	0.55	2.55	2.00	2.26	2.04	3.74	
YE and CA[c] n=14	mean					0.71	0.14	0.13	0.04	0.15	0.03	nd	Liu et al. 2003
	median					0.52	0.16	nd	nd	nd	nd	nd	
	min					0.10	nd	nd	nd	nd	nd	nd	
	max					2.48	0.27	0.93	0.25	1.23	0.16	0.01	
	SD					0.65	0.08	0.32	0.09	0.34	0.06	0.00	
	CV					0.92	0.54	2.54	2.07	2.27	2.04	3.74	

		Chlor-dane	CC	TC	TC/CC	Ald-rin	Diel-drin	Hepta-chlo	Heptac-hlor-Epoxide	Endo-sulfan I	Endosu-lfan II	Endri-n/Ketone	Ref.
YQB[d] n=6	mean	0.14	0.07	0.07	1.12								Yang et al. 2010
	median	0.14	0.07	0.07	1.07								
	min	0.06	0.03	0.01	0.20								
	max	0.20	0.10	0.10	2.33								
	SD	0.05	0.03	0.03	0.69								
	CV	0.37	0.38	0.46	0.62								
SMB[e] n=5	mean	0.10	0.06	0.04	0.71								
	median	0.11	0.07	0.04	0.71								
	min	0.07	0.04	0.03	0.50								
	max	0.12	0.08	0.05	1.00								
	SD	0.02	0.02	0.01	0.19								
	CV	0.20	0.27	0.20	0.27								

[a] YRE & AS is Yangtze River Estuary and adjacent Sea; [b, c] YE & CA is Yangtze Estuary and nearby Coastal Areas; [d] YQB is Yueqing Bay; [e] SMB is Sanmen Bay.

Table 4. Enantiomeric fractions (EFs) of chiral organochlorine pesticides in the YRE and the adjacent ECS surface sediments

		o,p'-DDD	*o,p'*-DDT	*α*-HCH	Ref.
YRE & AS[a] n=50	mean	0.38	0.51	0.42	Yang 2011
	median	0.38	0.51	0.42	
	min	0.27	0.34	0.34	
	max	0.49	0.73	0.50	
	SD	0.06	0.10	0.04	
	CV	0.15	0.19	0.09	
YQB[b] n=6	mean	0.22	0.12	0.39	Yang et al. 2010
	median	0.19	0.13	0.39	
	min	0.16	0.10	0.35	
	max	0.39	0.13	0.43	
	SD	0.09	0.01	0.03	
	CV	0.40	0.08	0.08	
SMB[c] n=5	mean	0.37	0.14	0.40	
	median	0.38	0.13	0.40	
	min	0.30	0.11	0.30	
	max	0.44	0.20	0.47	
	SD	0.07	0.04	0.06	
	CV	0.19	0.25	0.15	
XSB[d] n=8	mean	0.54	0.58	0.36	Yang et al. 2011
	median	0.54	0.58	0.36	
	min	0.50	0.50	0.31	
	max	0.56	0.63	0.41	
	SD	0.02	0.05	0.03	
	CV	0.04	0.08	0.10	

[a] YRE & AS is Yangtze River Estuary and adjacent Sea; [b] YQB is Yueqing Bay; [c] SMB is Sanmen Bay; [d] XSB is Xiangshan Bay.

Composition and Potential Source

The compositional characteristics and isomer ratios of HCHs can provide valuable information to trace potential sources of HCHs in the surface sediments. Technical HCH congers mixtures were widely used as commercial pesticides. These mixtures contained 60-70% *α*-HCH, 5-12% *β*-HCH, 10-12% *γ*-HCH, 6-10% *δ*-HCH, and 3-4% *ε*-HCH (Walker et al. 1999). *β*-HCH is the most stable HCH isomers, and *α*-HCH and *γ*-HCH can also be transformed into β-HCH (Walker et al. 1999; Willett et al. 1998). Therefore, the amount of β-HCH can be used to identify whether the HCH source is historical or recent (Law et al. 2001). As shown in Figure 3, β-HCH was the predominant isomer in the YRE & AS, YQB, SMB, XSB and ECS sediments, which accounted for 45-60% of the total HCHs in the surface sediments. This result indicated that the HCH residues in the YRE and the adjacent ECS sediments mainly derived from historical inputs. Furthermore, the ratio of α-HCH/γ-HCH is useful for differentiating the HCH sources from technical HCH or lindane usage. Technical HCH has a α-HCH/γ-HCH ratio of 5-7, and lindane has that ratio of nearly zero (Lee et al. 2001). γ-HCH is more easily degraded by microbes (Bebezet and Matsumara 1973) and it can be transferred via photolytic isomerization to α-HCH in the environment (Malaiyandi and Shah 1984). These γ-HCH properties result in an increasing α-HCH/γ-HCH ratio when technical HCH is released into the environment. Contrarily, decreasing ratio may result from the release of lindane. In these studies, the average α-HCH/γ-HCH ratios in the YRE & AS, YQB, SMB and XSB sediments were 6.48, 7.28, 7.18 and 8.74, respectively (Table 2). It suggested that the HCH residues in these areas predominately resulted from technical HCH. However, in the YE and CA sediments, the mean ratio was 0.01, and about 94% of the $\sum$HCHs were γ-HCH. These findings suggested recent lindane inputs mainly contributed to the accumulation of HCHs in the YE and CA sediments. Moreover, this pattern may be linked to coarse-grained sediments (Liu et al. 2003). In the YE sediments, the average α-HCH/γ-HCH ratio in the sediments was 4.74, and β-HCH only accounted for 7.18%. These results were associated with the new introduce of lindane.

Pollution of Other Organochlorine Pesticides in the YRE and the Adjacent ECS Surface Sediments

Residue Levels and Distribution Characteristics

The descriptive statistical analysis of other organochlorine pesticides, including chlordane, Aldrin, Dieldrin, Heptachlo, Heptachlor-Epoxide, Endosulfan Ⅰ, Endosulfan Ⅱ and Endrin/Ketone, that were detected in the YRE and the adjacent ECS surface sediments is given in Table 3. In the YRE & AS, YQB and SMB sediments, the chlordane levels varied from nd to 0.22

ng/g, with mean values from 0.10 to 0.14 ng/g. The other OCP concentrations were found between nd and 2.48 in the YE and CA sediments.

Composition and Potential Source

Potential sources of chlordane can be traced from its composition. *Trans*-chlordane (TC) and *cis*-chlordane (CC) are two chlordane isomers. The TC is preferentially degraded before the CC in environmental conditions. In technical chlordane products, the TC/CC ratio is about 1.17. Thus, fresh inputs of chlordane would result in a TC/CC ratio of 1.17 or greater. The TC/CC ratios in the YRE & AS sediments were from 0 to 5, with a mean value of 1.98. It suggested that the accumulation of chlordane presented in the sediments recently deposited. However, the mean ratios determined in the YQB and SMB sediments were < 1.17, implying the historical residues happened in these areas.

Ecotoxicological Risk

In the sediment quality assessment, the ERL and ERM were also employed to predict the potential impacts of chlordane, Dieldrin and Endrin. For chlordane, the ERL and ERM values are 0.5 and 6 ng/g, respectively (Long et al. 1990). Luckily, none of the sediments in the studies areas (YRE &AS, YQB and SMB) exceeded the ERL, implying a relatively safe environment regarding the presence of chlordane. The ERL value for Dieldrin and Endrin is 0.02 ng/g, and the ERM values are 8 and 45 ng/g, respectively (Long et al. 1990). In most of the YE and CA sediments, the Dieldrin concentrations were between the ERL and ERM except for two samples, which Dieldrin was not detected. Therefore, probable-effect would be occasionally caused by Dieldrin. Given Endrin was detected in only one sample and at a concentration much lower than the ERL, the adverse effects of Endrin were absent or minimal.

As one of the largest producers and users of OCPs, about 14 million hm^2 of farmlands were contaminated with OCPs in China. Large amount of OCPs from agricultural areas deposited into the Yangtze River sediments, which play a key role in the migration and redistribution of the contaminants. Therefore, additional measures should be taken by government organizations to manage and control the sediment quality in YRE and the adjacent ECS.

Table 5. Residue concentrations of PCBs in the YRE and the adjacent ECS surface sediments (ng/g)

		∑di-PCB	∑tri-PCB	∑tetra-PCB	∑penta-PCB	∑hexa-PCB	∑hepta-PCB	∑PCB	Ref.
YRE & AS[a] n=50	mean	1.42	5.14	2.34	1.06	0.16	0.09	10.22	Yang et al. 2012
	median	1.46	5.26	2.24	1.02	0.16	0.09	10.32	
	min	0.22	2.17	1.05	0.60	0.00	0.00	5.08	
	max	4.72	8.63	4.72	1.80	0.47	0.20	19.64	
	SD	0.93	1.75	0.88	0.28	0.09	0.04	3.16	
	CV	0.66	0.34	0.37	0.27	0.55	0.50	0.31	
YQB[b] n=9	mean	2.21	8.78	2.50	0.69	0.06	0.04	14.24	Yang et al. 2011
	median	2.30	9.09	2.50	0.61	0.07	0.04	14.62	
	min	1.68	6.40	1.36	0.35	0.02	0.03	9.80	
	max	2.54	9.99	3.83	1.38	0.10	0.05	17.77	
	SD	0.28	1.14	0.66	0.32	0.02	0.01	2.20	
	CV	0.13	0.13	0.27	0.46	0.35	0.18	0.15	
XSB[c] n=9	mean	1.88	7.12	1.66	0.70	0.05	0.03	11.40	Yang et al. 2011
	median	1.89	7.47	1.65	0.57	0.04	0.03	11.44	
	min	1.55	6.00	1.35	0.39	0.02	0.02	9.51	
	max	2.30	7.73	2.15	1.49	0.09	0.06	12.91	
	SD	0.23	0.73	0.27	0.36	0.02	0.01	1.24	
	CV	0.12	0.10	0.16	0.51	0.47	0.43	0.11	
SMB[d] n=12	mean	1.66	7.33	2.29	1.43	0.14	0.05	12.87	Yang et al. 2011
	median	1.69	7.07	2.20	1.21	0.12	0.03	12.41	
	min	1.13	5.19	1.58	0.67	0.09	0.02	9.30	
	max	2.33	11.05	3.81	2.65	0.31	0.26	19.60	
	SD	0.33	1.82	0.58	0.74	0.06	0.07	2.68	
	CV	0.20	0.25	0.25	0.52	0.44	1.29	0.21	
JJE[e] n=23	mean	1.82	1.92	21.36	2.07	1.48	0.45	29.08	Zhou et al. 2012
	median	1.77	1.53	6.85	1.39	0.98	0.27	12.48	
	min	0.64	0.86	1.40	0.13	0.01	0.01	4.93	
	max	2.85	7.44	100.08	10.91	8.21	1.71	108.79	
	SD	0.63	1.37	28.11	2.27	1.73	0.53	30.49	
	CV	0.34	0.71	1.32	1.10	1.17	1.17	1.05	
YE and CA[f] n=14	mean							2.67	Liu et al. 2004
	median							0.48	
	min							nd	
	max							18.95	
	SD							5.31	
	CV							1.99	

Table 5. (Continued)

		PCB1	PCB11	PCB29	PCB47	PCB 121	PCB 136	PCB 187	∑PCB	Ref.
YE and CA[g] n=14	mean	nd	0.11	2.54	nd	0.05	0.01	nd	2.70	Liu et al. 2003
	median	nd	nd	0.59	nd	nd	nd	nd	0.59	
	min	nd	nd	nd	nd	nd	nd	nd	nd	
	max	nd	0.65	17.79	nd	0.63	0.07	nd	18.95	
	SD	nd	0.21	4.98	nd	0.17	0.02	nd	5.29	
	CV	-	2.01	1.96	-	3.51	3.74	-	1.96	

[a] YRE & AS is Yangtze River Estuary and adjacent Sea; [b] YQB is Yueqing Bay; [c] XSB is Xiangshan Bay; [d] SMB is Sanmen Bay; [e] JJE is Jiaojiang Estuary; [f, g] YE and CA is Yangtze Estuary and nearby Coastal Areas.

ENANTIOMERIC FRACTIONS OF CHIRAL OCPS

Most OCPs are produced and used as a racemic mixture. When these OCPs are introduced into the environment, their enantiomers will be selectively degraded or transformed into their stereoisomer, depending on biological factors. Therefore, the enantiomeric fraction (EF) is a sensitive indicator to study for the fate of OCPs in different environmental media and can be used to trace their derivation. The α-HCH, *o,p'*-DDD, *o,p'*-DDT, TC and CC were well separated during the enantiomeric analysis of chiral OCPs in the of YRE and the adjacent ECS surface sediments, whereas only the EFs of the first three OCPs were detected (Table 4).

For *o,p'*-DDD, the EFs of all YRE & AS, YQB, and SMB samples were all < 0.5, implying a faster degradation for (-)-*o,p'*-DDD in the surface sediments from these areas. However, the selective *o,p'*-DDT degradation in the sediments was variable. In the YRE & AS and XSB sediments, most EF values were > 0.5, implying the enrichment of (+)-*o,p'*-DDT in these sediments. However, the EFs in the YQB and SMB sediments were all much lower than 0.5, which suggested that (+)-*o,p'*-DDT degraded faster in these sediments. Fluctuating *o,p'*-DDT degradation was also observed in other studies (Aigner et al. 1998). It can be explained that by the complex microbiological processes in the sediments, the enantioselectivity in degradation of chiral OCPs is also multiplex.

Since all of the samples had the EFs < 0.5 regarding to α-HCH, (+)-α-HCH were preferentially degraded and the accumulation of α-HCH was derived from historical applications in these sediments. The faster degradation of (+)-α-HCH than the other enantiomers was also found by Buser and Muller (Buser and Muller 1995) under anaerobic conditions. After several years of farming, these YRE and the adjacent ECS areas are rich in organic matters and present anaerobic condition, which could result in the enantioselective degradation of (+)-α-HCH. However, the enrichment of (+)-α-HCH was found in other environmental media, which revealed the complexity of the biological processes were responsible for OCP degradation (Moisey et al. 2001).

Pollution of PCBs in the YRE and the Adjacent ECS Surface Sediments

Residue Levels and Distribution Characteristics

Descriptive statistics of the PCB congeners with different number of Cl atoms in the YRE and the adjacent ECS surface sediments are listed in Table 4. PCBs were detected in all of the sediments that were sampled by Yang et al. (Yang et al. 2011, 2012) and Zhou et al. (2012), indicating ubiquitous contamination by PCBs in theses aquatic environment. Next, the total concentrations of all the individual congeners were calculated to determine the contamination levels of PCBs in the YRE and the adjacent ECS sediments. The total concentrations of PCBs were in the range from nd to 108.79 ng/g, with mean values from 2.67 to 29.08 ng/g. The greatest total PCB concentrations existed in Jiaojiang Estuary, which might be due to the PCB52 point sources that were associated with sewage discharged from neighboring industrial parks. Because of different sampling locations, PCB concentrations varied significantly in these studies. The highest concentration of PCBs also happened in the marginal filter, where 90% of the suspended matter and organic matter were deposited. It resulted in the accumulation of PCBs. The inner river of the Yangtze and Qiantang River had the second largest PCBs contamination due to their geographical characteristics. The higher levels of PCBs in the marginal filter and the inner river mouth then transported to the open sea, resulting the PCB concentration gradients. It was evidenced in the investigations by Zhou et al. (Zhou et al. 2012) and Yang et al. (Yang et al. 2011), who also found higher PCB levels in the inner part of the estuary. Moreover, the greater PCB contamination in the YQB compared to XSB and SMB sediments was related to the distance from the PCB sources. Wentai, Zhejiang province, released a great deal of PCBs during the disposal of used electronic appliances, and was potentially the main source of PCB pollution in the YQB sediments. In addition, the maximum of ∑PCBs concentration in the YE and CA sediments took place in the sites that were near sewage discharge outlets (Liu et al. 2003, 2004). Generally, the PCB pollution distribution was likely related to the random discharge of untreated contaminants from nearby sources. Additionally, except for the distance between the sampling sites and sources, hydrodynamic factors and the sediment organic matters textures potentially play an important role in the spatial distributions of PCBs in the sediments.

Composition and Potential Source

The homologue composition of PCBs in the YRE and the adjacent ECD surface sediments varied between different studies (Figure 4). In the case of the YRE & AS, YQB , XSB and SMB the proportions of tri-PCB were the largest (50.28 - 62.41%), followed by tetra-PCB (14.55-22.92%), di-PCB (12.91-16.52%), penta-PCB (4.81-11.07%), hexa-PCB (0.40-1.53%) and hepta-PCB (0.28-0.84%) (for XSB, the proportion of di-PCB was greater than that of tetra-PCB). The decreasing order of PCBs in Jiaojiang Estuary (JJE) was as follows: tetra-PCB > penta-PCB > tri-PCB > di-PCB > hexa-PCB > hepta-PCB. In this case, PCB52 contributed the most to the tetra-PCB enrichment in the sediments. In all the samples, the contents of octa-PCB, nona-PCB and deca-PCB were absent, and the concentrations of hexa-PCB and hepta-PCB were quite low or negligible. The most prevalent homologues were di-PCB, tri-PCB, tetra- PCB and penta-PCB, which totally accounted for 86.32 to 94.72% of the total PCBs. The phenomenon observed in these studies corresponded with the composition of the Chinese PCB when assuming #1 PCB (trichlorobiphenyl) and #2 PCB (pentachlorobiphenyl) were mixed at the ratio of 9:1 (Ren et al. 2007). It is noted that from 1965 to 1974, #1 PCB and #2 PCB were primarily used in power capacitors and paint additives in China (SEPA China 2003). Therefore, tri-PCB was the dominant congener produced in China, followed by tetra-PCB. This finding was in agreement with the contamination in Chinese rural soils, but not in urban soils. The main PCB pollution in urban soils in China was from hexa-PCB (31%) and tri-PCB (21%) (Ren et al. 2007).

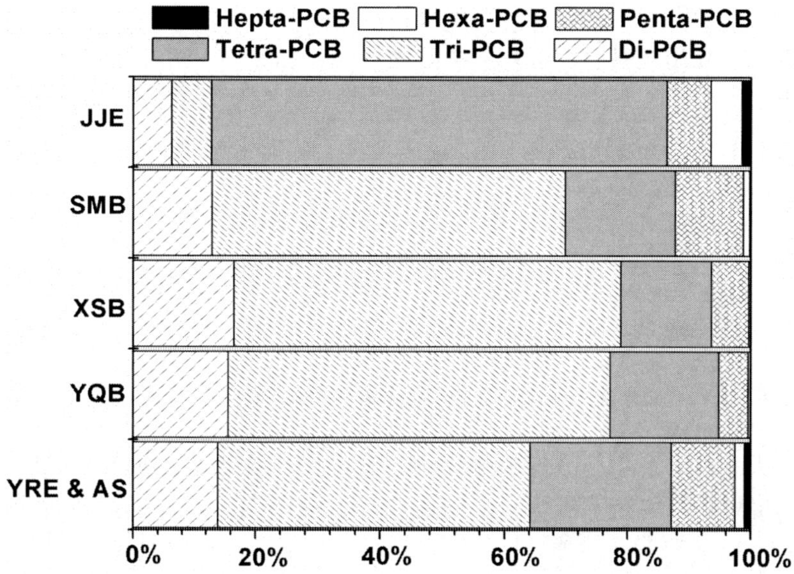

Figure 4. PCB composition in the surface sediments of Jiaojiang Estuary (JJE) (Zhou et al. 2012), Sanmen Bay (SMB) (Yang et al. 2011), Xiangshan Bay (XSB) (Yang et al. 2011), Yueqing Bay (YQB) (Yang et al. 2011) and Yangtze River Estuary and adjacent Sea (YRE & AS) (Yang 2011).

In the YRE & AS (Yang et al. 2012), an interesting result was found regarding the PCB congener profiles in the sediment at different longitudinal locations. In the western region, PCB congeners with low chlorine (di-, tri- and tetra- PCB) were found at higher levels, where large amount of organic matter and lower dissolved oxygen (DO) were the main properties of the aquatic environment after years of fish farming. Contrarily, DO reached a satuation of 100-104% in eastern region (China Bay Chi Compile Committee 1998). As a result, PCB dechlorination was preferred under anaerobic conditions in the more eutrophic West region than the open sea in the East.

Further information of PCB sources was obtained by comparing the PCB compositions of the sediment samples with principal commercial mixtures (Aroclor 1221, 1232, 1242, 1248, 1254 and 1260) by principal component analysis (PCA). PCA is a multivariate analysis to be adopted for understanding the interrelationships of a group of data by reducing the multidimensionality of variables between samples (Skrbic and Durisic-Mladenovic 2007). The PCA results revealed that Aroclor 1242 was the most common Aroclor used in the YRE and the adjacent ECS, which contained 47% tri-PCB, 30% tetra-PCB,

15% di-PCB and 8% penta-PCB on a weight basis. Whereas, the PCB pollution in samples located in JJE were mainly from the usage of Aroclor 1248, which is abundant with tetra-PCBs, and also with tri- and penta-PCBs. Although the production and direct application of PCBs was banned in the 1970s, they were still detected in the YRE and the adjacent EDS sediments. Their presence is caused by their continued use in electrical transformers, electrical capacitors, gas-transmission turbines and vacuum pumps. These PCBs are chronically released into the environment from landfills and sewage sludge incineration. Due to the arbitrary disposal of used electronics, Wentai, Zhejiang province may be a new source of PCBs for the YRE and the adjacent ECS surface sediments. The common use of Aroclor 1248 in the JJE potentially resulted from nearby industrial activities using chlorine compounds as raw materials or intermediates.

As regions with intense aquacultural and shipping activities, the YRE and the adjacent ECS are exposed to a variety of point and diffusion PCB sources. Thus, it is difficult to pinpoint a precise source-occurrence relationship. In addition, atmospheric precipitation, surface runoff and sediment re-suspension from tidal movement may contribute to the accumulation of PCBs in the YRE and the adjacent ECS surface sediment.

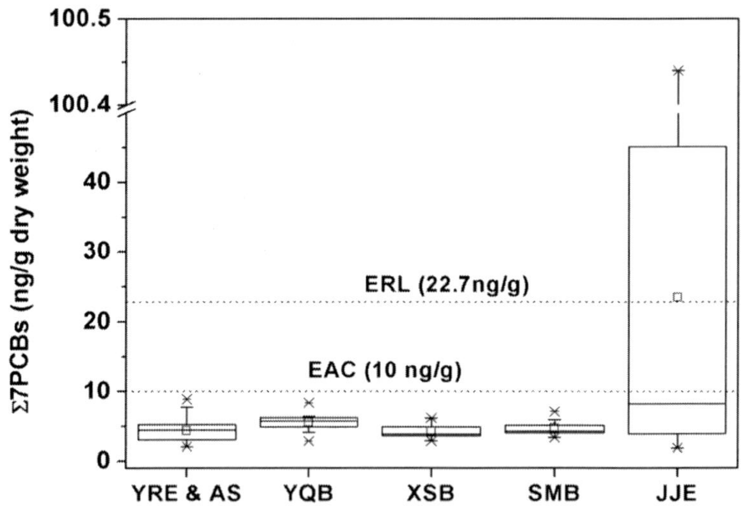

Figure 5. Ecotoxicological risk assessment of ∑7PCBs in the surface sediments of Yangtze River Estuary and adjacent Sea (YRE & AS) (Yang 2011), Yueqing Bay (YQB) (Yang et al. 2011), Xiangshan Bay (XSB) (Yang et al. 2011), Sanmen Bay (SMB) (Yang et al. 2011) and Jiaojiang Estuary (JJE) (Zhou et al. 2012).

Correlation with TOC

The accumulation and mobilization of PCBs can be affected by the physicochemical characteristics of sediments, such as sediment total organic matter. The concentrations of PCBs was significantly correlated with sediment TOC contents in the YQB, XSB and SMB (r = 0.83, 0.80 and 0.81, respectively). This finding revealed that TOC could be a good indicator for POPs pollution.

Ecotoxicological Risk

To evaluate the ecotoxicological significance of PCBs in sediments, the ecotoxicological assessment criteria (EAC) and the sediment quality guidelines (SQG) suggested by OSPAR (OSPAR Commission 2000) and the US-EPA/NOAA, respectively, were introduced. In the SQG, ERL and ERM are also specified. If one chemical concentration is below the ERL or EAC, the potential occurrence of adverse effects is minimal, while if one chemical concentration is above the ERM, adverse effects likely often happen. In this assessment, only seven PCB congers (i.e. PCB 28, 52, 101, 118, 138, 153 and 180) were calculated and compared with the guidelines since they could present all relevant degrees of chlorination. The total concentrations of the seven PCBs in all samples from the studied areas were below the ERM (180 ng/g) (Figure 5). In the YRE & AS, YQB, XSB and SMB surface sediments, none of the ∑7PCBs in surface sediment exceeded the EAC (10 ng/g). However, 43.48% of the samples from the JJE had ∑7PCBs concentrations beyond the EAC, and 34.78% of the sites had PCB concentrations above the ERL (22.7 ng/g). In addition, the average level of ∑7PCBs in the Jiaojiang Estuary sediments was greater than the EAC. It suggested that the JJE had a heavier PCB contamination than the other sites of the YRE and the adjacent ECS, which might pose potential adverse effects on the aquatic biota. Therefore, additional measures should be taken to amend the sediment contamination in the JJE. Moreover, the ecosystem condition at other sites could not be ignored, although the concentration of ∑7PCBs was below the criteria. More work is needed on aquatic biota to better understand the bioaccumulation of PCBs in the YRE and ECS surface sediments and their impacts on ecosystem and human health.

Conclusion

The residue levels of chlorinated organic contaminants in the YRE and the adjacent ECS surface sediments were presented in this chapter. A total of 115 and 117 samples with OCP (DDTs, HCHs and others) and PCB pollution, respectively, from the 10 previous studies were compiled and summarized. Based on their composition and a multivariate analysis method, the potential sources of OCPs and PCBs were traced. The concentrations of DDTs, HCHs and PCBs in the YRE and the adjacent ECS sediments ranged from nd to 33.10 ng/g, nd to 32.63 ng/g and nd to 108.79 ng/g, respectively. Location was the main factor influencing the spatial distribution of these pollutants. The marginal filter, inner river mouth and local sewage or wastewater discharge mainly contributed to the high levels of OCPs and PCBs in the sediments. The information obtained from the homologue composition and ratios of the parent chemicals and their metabolites could be used to determine whether the residues in sediments are from historical or recent use. In addition, this information can be used to determine which forms of OCPs and PCBs were attributed to their accumulation in sediments. It was evidenced that due to the complexity of the biological processes and the different sampling sites, the composition and sources of the chlorinated contaminants in the sediments varied. To estimate the impact of these contaminants on the ecosystem, ecotoxicological assessments were performed. Generally, most of the concentrations of OCPs and PCBs were in the low or intermediate range of the guidelines. It revealed the risk they posed to aquatic environment was mainly mineral or only happened occasionally. However, the p,p'-DDT concentration in the XSB and PCBs in JJE were high. Although the residue levels of OCPs and PCBs in the sediments were rather low or moderate, their long lifetime and biomagnification in sediment-dwelling organisms may greatly impact ecosystem and human health. Therefore, further work on the bioaccumulation of OCPs and PCBs in aquatic biota is necessary to obtain a holistic understanding of their risks. In addition, current guidelines and methods for environmental assessment are all based on the racemic compounds of chiral pesticides. The current guidelines may lead to an incorrect toxicological assessment due to the different biological properties of the enantiomers in sediments. Hereby, guidelines at enantiomeric levels for the potential risk assessment of chiral chemicals are needed to establish.

ACKNOWLEDGMENTS

This research was supported by the National Natural Science Foundation of China (Nos. 40973077 and 21177112).

REFERENCES

Aigner E. J., Leone A. D., & Falconer R. L. (1998a). Concentrations and enantiomeric ratios of organochlorine pesticides in soil from the US Corn Belt. *Environmental Science & Technology* 32 (9): 1162-1168.

Bebezet H. J., & Matsumara F. (1973). Isomerization of γ-HCH to α-HCH in the environment. *Nature* (243): 480-481.

Bettinetti R., Giarei C., & Provini A. (2003). Chemical analysis and sediment toxicity bioassays to assess the contamination of the River Lambro (Northern Italy). *Archives of Environmental Contamination and Toxicology* 45 (1): 72-78.

Bopp R. F., Simpson H. J., Olsen C. R., Trier R. M., & Kostyk N. (1982). Chlorinated hydrocarbons and radionuclide chronologies in sediments of the Hudson river and estuary, New-York. *Environmental Science & Techno logy* 16 (10): 666-676.

Breivik K., Sweetman A., Pacyna J. M., & Jones K. C. (2002). Towards a global historical emission inventory for selected PCB congeners - a mass balance approach 1. Global production and consumption. *Science of the Total Environment* 290 (1-3): 181-198.

Buser H. R. & Muller M. D. (1995). Isomer and enantioselective degradation of hexachlorocyclohexane isomers in sewage-sludge under anaerobic conditions. *Environmental Science & Technology* 29 (3): 664-672.

Chen L., Yu C. N., Shen C. F., Zhang C. K., Liu L., Shen K. L., Tang X. J., & Chen Y. X. (2010). Study on adverse impact of e-waste disassembly on surface sediment in East China by chemical analysis and bioassays. *Journal of Soils And Sediments* 10 (3): 359-367.

China Bay Chi Compile Committee. (1998). *China Bay Chi Ocean press,* Beijing. pp: 206-208 (in Chinese).

Daskalakis K. D. & Oconnor T. P. (1995). Distribution of chemical concentrations in us coastal and estuarine sediment. *Marine Environmental Research* 40 (4): 381-398.

de Brito A., Ueno D., Takahashi S. & Tanabe S. (2002). Organochlorine and butyltin residues in walleye pollock (Theragra chalcogramma) from Bering Sea, Gulf of Alaska and Japan Sea. *Chemosphere* 46 (3): 401-411.

Doong R. A., Peng C. K., Sun Y. C. & Liao P. L. (2002). Composition and distribution of organochlorine pesticide residues in surface sediments from the Wu-Shi River estuary, Taiwan. *Marine Pollution Bulletin* 45 (1-12SI): 246-253.

Feng H., Cochran J. K., Lwiza H., Brownawell B. J. & Hirschberg D. J. (1998). Distribution of heavy metal and PCB contaminants in the sediments of an urban estuary: The Hudson River. *Marine Environmental Research* 45 (1): 69-88.

Genualdi S. A., Simonich S., Primbs T. K., Bidleman T. F., Jantunen L. M., Ryoo K. S. & Zhu T. (2009). Enantiomeric Signatures of Organochlorine Pesticides in Asian, Trans-Pacific, and Western US Air Masses. *Environmental Science & Technology* 43 (8): 2806-2811.

Han W. L., Feng J. L., Gu Z. P., Wu M. H., Sheng G. Y. & Fu J. M. (2010). Polychlorinated biphenyls in the atmosphere of Taizhou, a major e-waste dismantling area in China. *Journal of Environmental Sciences-China* 22 (4): 589-597.

Hitch R. K., & Day H. R. (1992). Unusual persistence of DDT in some western USA soils. *Bulletin of Environmental Contamination and Toxicology* 48 (2): 259-264.

Kalantzi O. I., Alcock R. E., Johnston P. A., Santillo D., Stringer R. L., Thomas G. O., & Jones K. C. (2001). The global distribution of PCBs and organochlorine pesticides in butter. *Environmental Science & Technology* 35 (6): 1013-1018.

Law S. A., Diamond M. L., Helm P. A., Jantunen L. M. & Alaee M. (2001). Factors affecting the occurrence and enantiomeric degradation of hexachlorocyclohexane isomers in northern and temperate aquatic systems. *Environmental Toxicology and Chemistry* 20 (12): 2690-2698.

Lee K. T., Tanabe S. & Koh C. H. (2001). Distribution of organochlorine pesticides in sediments from Kyeonggi Bay and nearby areas, Korea. *Environmental Pollution* 114 (2): 207-213.

Lin T., Hu L. M., Shi X. F., Li Y. Y., Guo Z. G. & Zhang G. (2012). Distribution and sources of organochlorine pesticides in sediments of the coastal East China Sea. *Marine Pollution Bulletin* 64 (8): 1549-1555.

Liu M., Cheng S. B., Ou D. N., Yang Y., Liu H. L., Hou L. J., Gao L. & Xu S. Y. (2008). Organochlorine pesticides in surface sediments and suspended particulate matters from the Yangtze estuary, China. *Environmental Pollution* 156 (1): 168-173.

Liu M., Yang Y., Hou L., Xu S., Ou D., Zhang B. & Liu Q. (2003). Chlorinated organic contaminants in surface sediments from the Yangtze Estuary and nearby coastal areas, China. *Marine Pollution Bulletin* 46 (5): 672-676.

Liu M., Yang Y., Xu S., Hou L., Liu Q., Ou D. & Jiang H. (2004). Persistent organic pollutants (POPS) in intertidal surface sediments from the Yangtze Estuarine and coastal areas, China. *Journal of Coastal Research* (43): 162-170.

Long E. R., Macdonald D. D., Smith S. L. & Calder F. D. (1995). Incidence of adverse biological effects within ranges of chemical concentrations in marine and estuarine sediments. *Environmental Management* 19 (1): 81-97.

Long E. R. & Morgan L. G. (1990). *The potential for biological effects of sediment-sorbed contaminants tested in the National Status and Trends Program. NOAA Tech.* Memo. NOS OMA 52. US National Oceanic and Atmospheric Administration, Seattle, Washington, 175 pp.

Malaiyandi M., & Shah S. M. (1984). Evidence of photoisomerization of hexachlorocyclohexane isomers in the ecosphere. *Journal Of Environmental Science and Health Part A-Environmental Science and Engineering & Toxic and Hazardous Substance Control* 19 (8): 887-910.

Miller S. (1982). The persistent PCB problem. *Environmental Science & Technology* 16 (2): A98-A99.

Moisey J., Fisk A. T., Hobson K. A. & Norstrom R. J. (2001). Hexachlorocyclohexane (HCH) isomers and chiral signatures of alpha-HCH in the arctic marine food web of the Northwater Polynya. *Environmental Science & Technology* 35 (10): 1920-1927.

Nemirovskaya I. A. (2009). Transformation of hydrocarbons in the marginal filter of the Ob River. *Doklady Earth Sciences* 429 (2): 1536-1541.

OSPAR Commission. (2000). *Quality Status Report 2000*. OSPAR Commission, London, pp 41-63.

PedersenBjergaard S., Semb S. I., Vedde J., Brevik E. M. & Greibrokk T. (1996). Comparison of GC-ECD, GC-MS and GC-AED for the determination of polychlorinated biphenyls in highly contaminated marine sediments. *Chromatographia* 43 (1-2): 44-52.

Qiu X. H., Zhu T., Yao B., Hu J. X. & Hu S. W. (2005). Contribution of dicofol to the current DDT pollution in China. *Environmental Science & Technology* 39 (12): 4385-4390.

Ren N. Q., Que M. X., Li Y. F., Liu Y., Wan X. N., Xu D. D., Sverko E. & Ma J. M. (2007). Polychlorinated biphenyls in chinese surface soils. *Environmental Science & Technology* 41 (11): 3871-3876.

SEPA China. (2003). *Buiding the Capacity of the People's Republic of China to Implement the Stockholm Convention on POPs and Develop a National Implementation Plan.* In: GEF Project Brief (GF/CPR/02/010). China State Environmental Protection Administration, Beijing.

Shen C. F., Huang S. B., Wang Z. J., Qiao M., Tang X. J., Yu C. N., Shi D. Z., Zhu Y. F., Shi J. Y., Chen X. C., Setty K. & Chen Y. X. (2008). Identification of Ah receptor agonists in soil of E-waste recycling sites from Taizhou area in China. *Environmental Science & Technology* 42 (1): 49-55.

Skrbic B. & Durisic-Mladenovic N. (2007). Principal component analysis for soil contamination with organochlorine compounds. *Chemosphere* 68 (11): 2144-2152.

Tao S., Liu W. X., Li Y., Yang Y., Zuo Q., Li B. G. & Cao J. (2008). Organochlorine Pesticides Contaminated Surface Soil as Reemission Source in the Haihe Plain, China. *Environmental Science & Technology* 42 (22): 8395-8400.

Totten L. A., Panangadan M., Eisenreich S. I., Cavallo G. J. & Fikslin T. J. (2006). Direct and indirect atmospheric deposition of PCBs to the Delaware river watershed. *Environmental Science & Technology* 40 (7): 2171-2176.

Walker K., Vallero D. A. & Lewis R. G. (1999). Factors influencing the distribution of lindane and other hexachlorocyclohexanes in the environ ment. *Environmental Science & Technology* 33 (24): 4373-4378.

Wang T. Y., Lu Y. L., Zhang H. & Shi Y. J. (2005). Contamination of persistent organic pollutants (POPs) and relevant management in China. *Environment International* 31 (6): 813-821.

Willett K. L., Ulrich E. M. & Hites R. A. (1998). Differential toxicity and environmental fates of hexachlorocyclohexane isomers. *Environmental Science & Technology* 32 (15): 2197-2207.

Wu H. L., Jia H. L., Zhou L., Yang M. & Li Y. F. (2007). The production, use, residual level and analysis of organochlorine pesticides (HCH, DDT) in Taihu Lake. <http:// www.china-pops.net/yjnew_view.asp?id=31> (in Chinese).

Yamashita N., Kannan K., Imagawa T., Villeneuve D. L., Hashimoto S., Miyazaki A. & Giesy J. P. (2000). Vertical profile of polychlorinated dibenzo-p-dioxins, dibenzofurans, naphthalenes, biphenyls, polycyclic aromatic hydrocarbons, and alkylphenols in a sediment core from Tokyo Bay, Japan. *Environmental Science & Technology* 34 (17): 3560-3567.

Yang H. Y. (2011). *Study on organochlorine pesticides and polychlorinated biphenyls in Yangtze River Delta and adjacent sea.* Zhejiang University of Technology. (in Chinese)

Yang H. Y., Xue B., Jin L. X., Zhou S. S. & Liu W. P. (2011). Polychlorinated biphenyls in surface sediments of Yueqing Bay, Xiangshan Bay, and Sanmen Bay in East China Sea. *Chemosphere* 83 (2): 137-143.

Yang H. Y., Xue B., Yu P., Zhou S. S. & Liu W. P. 2010. Residues and enantiomeric profiling of organochlorine pesticides in sediments from Yueqing Bay and Sanmen Bay, East China Sea. *Chemosphere* 80 (6): 652-659.

Yang H. Y., Zhuo S. S., Xue B., Zhang C. L. & Liu W. P. (2012). Distribution, historical trends and inventories of polychlorinated biphenyls in sediments from Yangtze River Estuary and adjacent East China Sea. *Environmental Pollution* 169: 20-26.

Yang R. Q., Jiang G B., Zhou Q. F., Yuan C. G. & Shi J. B. (2005). Occurrence and distribution of organochlorine pesticides (HCH and DDT) in sediments collected from East China Sea. *Environment International* 31 (6): 799-804.

Yang Y. H., Zhou S. S., Li Y. Y., Xue B. & Liu T. (2011). Residues and chiral signatures of organochlorine pesticides in sediments from Xiangshan Bay, East China Sea. *Journal Of Environmental Science and Health Part B-Pesticides Food Contaminants and Agricultural Wastes* 46 (2): 105-111.

Yang Z. F., Shen Z. Y., Gao F., Tang Z. W. & Niu J. F. (2009). Occurrence and possible sources of polychlorinated biphenyls in surface sediments from the Wuhan reach of the Yangtze River, China. *Chemosphere* 74 (11): 1522-1530.

Zhang G., Li J., Cheng H., Li X., Xu W. & Jones K. C. (2007). Distribution of organochlorine pesticides in the northern South China Sea: Implications for land outflow and air-sea exchange. *Environmental Science & Technology* 41 (11): 3884-3890.

Zhang G., Parker A., House A., Mai B. X. Li X. D., Kang Y. H., & Wang Z. S. (2002). Sedimentary records of DDT and HCH in the Pearl River Delta, South China. *Environmental Science & Technology* 36 (17): 3671-3677.

Zhang Z., Liu L. Y., Li Y. F., Wang D. G., Jia H. L., Harner T., Sverko E., Wan X. N., Xu D. D., Ren N. Q., Ma J. M. & Pozo, K. (2008). Analysis of polychlorinated biphenyls in concurrently sampled Chinese air and surface soil. *Environmental Science & Technology* 42 (17): 6514-6518.

Zhou J. L., Maskaoui K., Qiu Y. W., Hong H. S. & Wang Z. D. (2001). Polychlorinated biphenyl congeners and organochlorine insecticides in the water column and sediments of Daya Bay, China. *Environmental Pollution* 113 (3): 373-384.

Zhou S. S., Shao L. Y., Yang H. Y., Wang C. & Liu W. P. (2012). Residues and sources recognition of polychlorinated biphenyls in surface sediments of Jiaojiang Estuary, East China Sea. *Marine Pollution Bulletin* 64 (3): 539-545.

In: Yangtze River ISBN: 978-1-62618-285-1
Editors: T. B. Maloney, B. R. Hutchins

Chapter 2

ENVIRONMENTAL AND LAND-USE CHANGES IN THE TIBETAN PLATEAU SECTION OF THE UPPER YANGTZE RIVER BASIN DURING THE LAST 50 YEARS

Stuart A. Harris*

Department of Geography, University of Calgary, Calgary, Alberta, Canada

ABSTRACT

The upper Yangtze River basin can be divided into a western section including the Qinghai-Tibetan Plateau and the eastern slopes down to Yibin City, and the eastern region above the Three Gorges Dam. This chapter discusses the environmental and land-use changes that have taken place during the last 50 years in the western portion. The Yangtze River flows from the terminus of a glacier at 6,500 m on the Tanggula Range, crossing the northwest part of the Qinghai-Tibetan Plateau at an elevation between 4700 m and 3800 m. This is a cold, arid area with a semi-desert vegetation, currently undergoing desertification due to climate change and recent land-use practices. The river then descends slowly across cold high-altitude *Kobresia* meadows before plunging down to about 300 m elevation through spectacular gorges with a mixed coniferous forest vegetation. Exploitation of forests, mining and construction of dams has and is modifying these gorges, though several substantial areas are now

* E-mail: harriss@ucalgary.ca.

protected. Reforestation of mountain sides is being conducted to try to revegetate the steep slopes and reduce erosion. Mining is now restricted to specific areas, and there are resettlement plans for the people displaced by the proposed construction of a sequence of cascading dams planned to provide much needed electricity to meet the demands of an increasingly industrialized nation.

Introduction

As the third longest river in the world (6236 km), the Yangtze River is the longest river in Asia with a catchment area of approximately 1.81 million square kilometers. It is usually treated as consisting of three contrasting basins, of which the Yangtze Basin drains an area of about 1 million square kilometers from its source at the terminus of a glacier at about 6,500m a.m.s.l. on the south side of Mount Gelangdang in the Tanggula Range of the Qinghai-Tibet Plateau (QTP), east to Yichang, 40 km downstream of the Three Gorges dam. The Upper Basin can be divided into three separate parts, viz., from the source down to Yibin City, Sichuan Province, China, at about 300m a.m.s.l., the Sichuan Basin and surrounding mountains, and the three mountain chains that produce the spectacular Three Gorges (Figure 1). This Chapter will examine the western portion consisting of the QTP and the river course down the steep eastern slopes of the Plateau. These have significantly different environments and problems to those of the central and eastern sections of the Upper Yangtze River.

Close to its source, the Yangtze River traverses a large plateau area with alpine cushion plants, cold dry steppe and meadows, underlain by continuous and discontinuous permafrost. Eastwards it passes through grassy steppes into the Sichuan Basin evergreen broadleaf forests on the eastern slopes of the Tibet Plateau. The major concern on the Plateau is to minimize it passes through grassy steppes into the Sichuan Basin evergreen broadleaf forests on the eastern slopes of the Tibet Plateau. The major concern on the Plateau is to minimize desertification in the very dry, cold climate. Over the last 50 years, both precipitation and air temperature have been changing on the QTP, resulting in substantial environmental changes. The area has been modified by the influx of Tibetan herders, construction of a product oil pipeline, highways, railways and nowFreeways from Golmud in West Qinghai Province to Lhasa in Central Tibet, and from Xi'ning to Yushu (Gyệgu) in Southern Qinghai Province on northeastern QTP. There are also plans for water diversions from the upper Yangtze River into the upper reaches of the Yellow River, as well as other changes in land use and grazing. Both the population and the vegetation have been changing, permafrost has been degrading, and there have been significant changes to the hydrology and hydrogeology of the area.

The eastern slopes below about 4000m have abundant precipitation from the East Asian and Indian Monsoons, but these are shallow air masses that rarely extend upslope beyond about 4600m onto the lower slopes of the arid QTP. In this area, the problems are slope instability, deforestation and the proposed development of cascading dams to produce hydroelectric power.

This Chapter will examine the substantial changes in the environment and land use that have affected these two areas during the last 50 years, as well as the difficult decisions that have been made regarding both use and protection of the area.

THE PHYSICAL ENVIRONMENT

1. Topography

With an average elevation of about 4,500 m, the Tibetan Plateau is often referred to as "the roof of the world". The highest corner is in the north-west, and from the Plateau, 7 major streams flow southwestwards (Indus River), southwards (Ganges-Brahmaputra-Meghna, Lancang-Mekong, Irrawaddy and Nu-Salween Rivers), eastwards (Yangtze River), or northeastwards (Yellow River) to the sea (Varis et al., 2012). They supply water to a vast, densely populated area of Southeast Asia and the South Asia Subcontinent which has a steadily increasing population. The rivers have been extensively modified over the past few millennia, but only the Indus and the Yellow River suffer from a shortage of water. The longest among them is the Yangtze River, the lower reaches of which have been the cradle of Chinese civilization at least since Neolithic times, about 8,000 B.C. (Ren, 1996).

The Tibetan Plateau is located north of the Himalayan Mountains, and has experienced a rapid uplift of ca. 3000m in the past 2 Ma (Xu et al., 2000). This has resulted in rapid incision of the Yangtze River and its tributaries into the rocks of the eastern slopes of Tibet. This is estimated to have removed 5km thickness since the Eocene (Wilson and Fowler, 2011; Kirby and Ouimet, 2012). Originally the streams flowing eastwards from the Plateau turned south into the Indian Ocean, while the Yangtze River only drained the lower Yangtze Basin. During the Cenozoic Era, the river extended westwards capturing the streams flowing southeastwards from the Tibetan Plateau in the vicinity of the Tiger Leaping Gorge (Hu'tiao'xia), 25 km northwest of Lijiang (Figure 2), thus producing the spectacular bends of the river valley in that area. This has resulted in very steep gradients along the river to the northwest of the gorge along the eastern margin of the Plateau. The river plunges through these deep gorges which separate the rolling eastern slopes of the plateau above 3500m to the west from the relatively low-elevational area northwest of Lijiang.

This capture also provided sufficient flow to allow the incision to keep pace with the uplift of the rising north-south trending fault blocks along the Three Gorges section of the middle Yangtze Basin (Richardson et al., 2010) during the last 0.75 million years (Fang et al., 2006), thus producing the famous Gorges. The Sichuan Basin was not uplifted to the same extent. On the eastern QTP, the rolling eastern slopes descend rapidly from the rolling QTP which lies above about 4500m. The plateau is traversed by substantial east-west trending mountain ranges, e.g., the Tanggula Shan and the Kunlun Shan. Sand dunes and deflation hollows are common along the eastern part of this plateau, while shallow thermokarst lakes are widespread at the higher elevations. The mountain ranges exhibit cold-based glaciers in their valleys, and it is from one of these that the river originates. East of the QTP, the river flows east through Yushu (Gyệgu) Tobetan Prefecture, Qinghai Province which was the epicenter (source at 10km depth) of the April 14th, 2010 earthquake that devastated that City (6.9 on the Richter scale).

In 2008, a major earthquake also occurred along the Yangtze River in Sichuan Province. All the faults along the Yangtze River are active, with both horizontal and vertical movements occurring frequently. On the QTP itself, fault scarps are commonly seen traversing alluvial fans, one, over 1000 km long was produced by the Kunlun Earthquake along the south side of the Kunlun Shan on 14th November, 2011.

2. Climate

The climate of the catchment ranges from that of high mountains with glaciers ranging up to above 7000 m a.m.s.l. (e.g., the Gongga Summit is at 7556 m), down to the hot, wet monsoon lands in the Sichuan Basin. Four air masses affect the area, viz., the cA/cP air of the Siberian high from the north in winter, the East China summer (Pacific) Monsoon from the east, the Indian Monsoon that swings northwards around the east end of the Himalayas, and the rather dry continental temperate air coming from the west. The latter alternates with the cA/cP air in winter, while the monsoons bring heavy precipitation to the eastern slopes of the QTP in summer. Precipitation on the Plateau is limited to incursions of small amounts of precipitation in summer from the Indian monsoon, primarily affecting the southeast slopes.

The China Meteorological Administration has measured the temperature and precipitation at a network of stations across the Yangtze River catchment area since 1961. The results indicate that the extreme values for mean annual air temperature (MAAT) decrease from a maximum of about 22°C at just over 2000m a.m.s.l. to a minimum of just under 6°C at about 4650m a.m.s.l. The whole data set produces an apparent lapse rate of -4.8°C/km (Wang et al., 2012). In July, the lapse rate is only -4.3°C/km, whereas in January, it is - 5.0/km. There is a marked difference in the January lapse rates between the stations located north and south of 53°N latitude, indicating that the cA/cP air only affects the lower altitudes of the eastern slopes of the Plateau as far south as that latitude, although the projected January lapse rates meet above about 5000m elevation. Unfortunately, there are no long-term weather stations above about 4800m a.m.s.l., so there is no data on the climate along the mountain chains. Wang et al. (2012) concluded that the variation in MAAT increases at high altitudes. Local cold air drainage probably accounts for a lot of the variation in MAAT in the depressions and valleys on the QTP, though this has not been examined to date.

There is a general consensus amongst Chinese authors that there has been an obvious change in climate over the eastern QTP in the last 50 years. Thus Wang et al. (2012) concluded that there was an increase in warming of the near-surface lapse rate due to altitude of about 1.0°C/km/100a. Jin et al. (2012) reported that during the period 1961-2005, the MAAT rose by an average of 1.12°C, an average rate of 0.025°C/a. The change was greatest during the period 1995-2006 in areas with only seasonal frost (Wang, C., personal communication, November, 2011), but this recent rate of change is not indicated in the data from weather stations in areas with permafrost (Jin et al., 2012). Li et al. (2010) produced maps of the rate of change in MAAT for the eastern QTP for 1961-2007 which show that there is actually considerable spatial variation with some areas actually cooling. However, the glaciers are retreating, permafrost is degrading (see below), and the current small increase in water yield will disappear when the thawing ice has been melted (Chen et al., 2001).

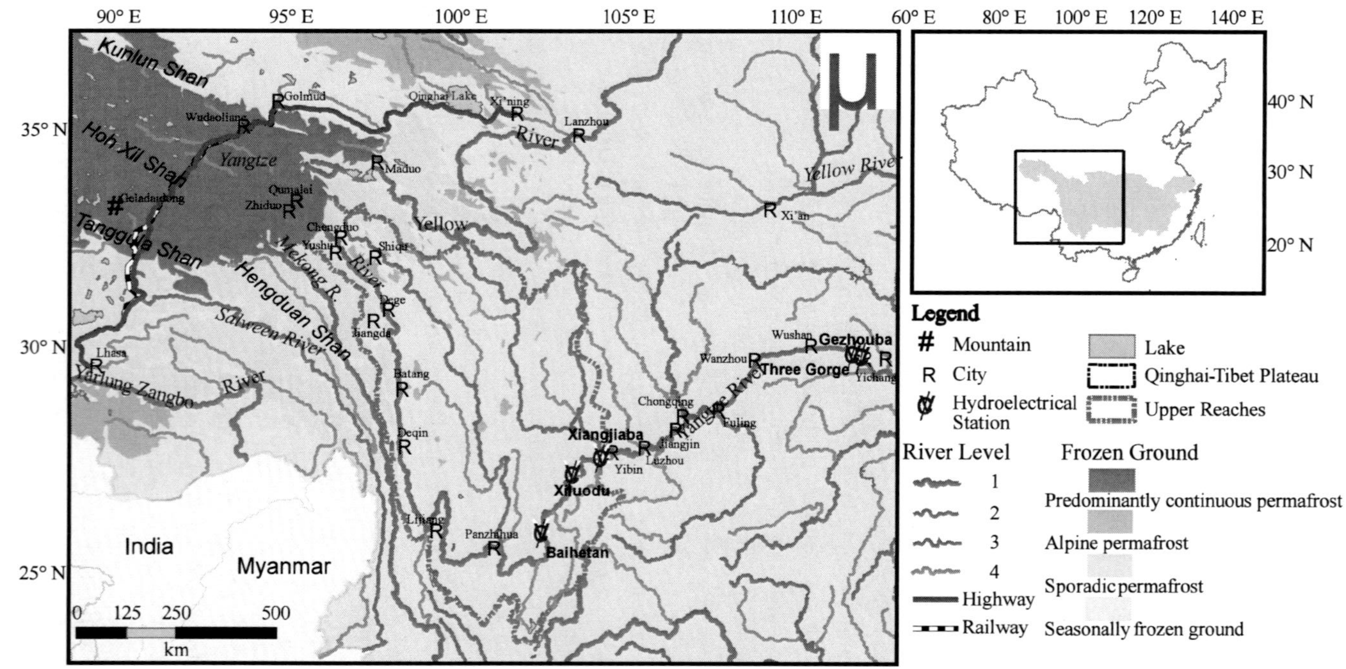

Figure 1. Map of the Yangtze River showing the western part of the Upper Drainage Basin.

On the high altitude desert, the mean annual precipitation can be as low as 20mm/a. This contrasts with the precipitation of 400-700mm/a, with a frost-free season of 20-100 days in the high-cold meadow area (Chang, 1981), though Jin et al (2009) report 15-20 frost-free days/a in the area of the adjacent Yellow River headwaters which has been studied more intensively (Qinghai Geological Survey, 2002; Ding et al., 2006).

In the high-cold meadow area, summer thunderstorms with hail may occur, and only 10-20% of the precipitation occurs during the seven cold months. Solar radiation is strong with total annual sunshine hours ranging from 2500-2800h, while mean annual insolation varies from 6400 to 6800 MJ m^{-2}. There is evidence of increasing winter snow depth since the late 1970's (Xin et al., 2010, Figure 1), based on the average annual snow depth data from all the weather stations. Average annual evaporation is about 1000-1500mm, though there was an average decrease in total precipitation of $25mm/a^{-1}$ between 1980 and 2000 (Ding et al., 2006).

The data for change in mean annual precipitation across the eastern QTP for 1961-2007 indicate a marked average increase in both the variability from year to year and in total precipitation after about 1975 (Li et al., 2010, Figure 4). This increase is not universal, but tends to be concentrated around certain locations along the boundary of the penetration of the monsoon rains on the eastern margin of the Plateau. Westwards and northwards, there appears to be a small decrease.

Chen et al. (2006) examined time series (1961-2000) for Penman-Monteith potential evapotranspiration estimates for 101 weather stations on the QTP and found that it decreased over time in all seasons. The overall average decrease was 13.1%/10a, or 2% of the total. The biggest change is in winter and spring (at 80% of all stations). An increase in evapotranspiration was only in the desert of the Qaidam Basin, well to the north of the Yangtze Basin. The decreases were correlated with an increase in wind speed and to a limited extent, relative humidity, thus suggesting a general increase in the influence of the monsoon circulation over the eastern QTP. The authors discounted the effects of increasing MAAT.

There are also changes in climate occurring in the mixed coniferous forest on the eastern slopes of the QTP. Lü and Lü (2002) report that the annual precipitation has decreased by 23.5-28.6 mm/a over a period of 10 years. The subalpine forest zone between 2600 m and 3600m has undergone an average increase in mean annual temperature of 0.19-0.25°C/10years over the last 30 years (Liu and Hou, 1998; Tan et al., 2000).

Figure 2. Tiger Leaping Gorge viewed from upstream.

3. Vegetation

The vegetation on the QTP in the catchment of the Yangtze River ranges from high-cold semidesert (Figure 3) with a mean annual precipitation below 100 mm/a in the west of the area, through high-cold steppe to high-cold alpine meadow in the east (Chang, 1981). The transition from steppe to meadow occurs at about the 400 mm/a isohyet in the north, increasing to 500 mm/a in the south. The species present in the catchment area of the Yangtze River are a mixture of Tibetan endemics, e.g., *Sausurea* and *Leontopodium* species with *Androsaceae tapete*, and Central Asiatic plants, e.g., *Axyris, Ceratoides, Chenopodium, Kochia,* and *Suaeda* species that are found in Mongolia and Kazahkstan (Wu, 1985-1987). At 4000m, the mean July air temperature is about 11°C. The westerly winds bring negligible snow in winter, so there is no insulating snow cover to protect the plants from the cold air.

The Tibetan endemic cushion plants above about 4400m are particularly interesting since they are still evolving as uplift of the plateau continues. They usually occur in specific altitudinal zones, often only extending about 300m in altitudinal range. Others are more widely distributed throughout the steppe zone. Chang (1981) argued for the distribution being a "high mountain zonation" in place of "alpine vegetation". Actually, a similar but less marked zonation is found on the mountains in North America which have a much lower elevation, so that this is probably a characteristic of natural mountain vegetation elsewhere in the world. The situation in the Alps is severely modified by destruction of the original forests and by grazing, so the vegetation there is not typical of relatively undisturbed vegetation zonation on mountains elsewhere in the world.

The vegetation of the alpine meadows is often composed of *Kobresia* species (Figure 4), while *Carex moorcroftii* is widespread on the sandy soils and dunes. Many species of *Androsaceae, Astragalus, Gentiana, Oxytropis, Potentilla, Rheum* and *Saussurea* occur in the zone of cushion plants and cold steppe. Li and Wu (1991) discuss the origins of the Tibetan flora, concluding that it ranks as the Qinghai-Xizang Plateau subkingdom, with some endemic families and genera from the Central Asia region and the Mediterranean region. It is completely different to the flora of Afghanistan and the region south of the Himalayas (Breckle and Rafiqpoor, 2010).

The major challenge on the QTP is how to minimize desertification in spite of grazing, increasing populations of herders, and the consequences of the anthropogenic changes to the landscape caused by the Qinghai-Tibet Transportation Corridor (Figure 3). The latter involves pipelines, roads, military depots and settlements to support and maintain these facilities.

At lower elevations to the east, mountain coniferous forests characterized by *Picea* and *Pinus* species appear (Figure 5), grading down into mixed evergreen forest with oaks, rhododendrons, maples and bamboos, together with a rich, diverse assemblage of forbs. This vegetation is dependent on the hot humid conditions in summer associated with the summer monsoon rains. The upper limit of this vegetation indicates the upper limit of regular monsoon conditions, though the summer rains do reach the meadows at irregular intervals, providing sufficient moisture for the *Kobresia* meadows.

Figure 3. Wind-eroded steppe-meadow due to overgrazing, pika burrows and contraction cracking.

Figure 4. *Kobresia* meadow on the eastern slopes of the QTP, with eroded steppe-meadow on the hills in the distance.

Figure 5. Mountain coniferous forest west of Lijiang.

Figure 6. Herd of Tibetan Antelope on the eroded area along the Golmud-Lhasa Highway. In the background is typical, partly eroded alpine-steppe.

4. Wildlife

China has 10% of the world's animal and plant species. Both the QTP and its forested eastern slopes are where a number of endangered or rare animals have survived. The area is vast and remote, with limited accessibility. The high elevation and lack of roads on the QTP prevents most visitors from wandering about, so this is a refuge for a number of rare species. Among these are the following mammals; Wolf, Red Fox, Sand Fox, Brown Bear, Otter, Alpine Weasel, Steppe Polecat, Eurasian Badger, Pallas's Cat, Lynx, Snow Leopard, Wild Ass, Alpine Musk Deer, Wild Yak, Tibetan Antelope (Figure 6), Tibetan Gazelle, White-lipped Deer, Blue Sheep, Tibetan Argali, Himalayan Marmot, Tibetan Hamster, Black-lipped Pika, Large-eared Pika, Glovers Pika and Wooly Hare (Plateau Perspectives, no date).

On the eastern slopes of the QTP in the remote parts of the mixed coniferous forests are found the Giant Panda, White-lipped Deer, Golden Monkeys, Gnu, and the Red Panda The Giant Panda is very slow moving and was used for food by humans since at least 10,000 years ago. Its survival in these mountains indicates how few people venture into the more remote parts.

Figure 7. Wind-eroded landscape on the upper eastern slopes of the QTP.

Soils

The soils of the QTP consist of shallow, base-rich, neutral or slightly alkaline (pH 6-8.5) alpine and subalpine "meadow soils" (Institute of Soil Science, 1986). In many places they are severely eroded by both wind and sheet wash except where there is a continuous vegetation mat. Such places are usually in depressions with a *Kobresia* cover. Saline soils are reasonably common on the lower lying parts of the QTP due to the intense evaporation of surface waters in the arid climate. They mainly occur at lower elevations (4300-4700m a.m.s.l.). Soil erosion is primarily by wind or by sheet wash on the finer textured soils at this elevation, and tends to smooth out the landscape. Wind action has produced sizeable areas of sand dunes especially east of the Golmud-Lhasa transportation corridor (Figure 7). These processes are facilitated by the sparse vegetation cover on all but the wettest areas.

Below about 3500m a.m.s.l. on the upper slopes of the east side of the QTP, they grade into Cinnamon soils along the lower slopes of the valleys and gorges. Since the mean annual precipitation is much higher, these are generally somewhat acid (pH 5-6), have a higher organic matter content (c. 2-3%), and have a lower base status. On the steep mountain slopes, the soils are usually very shallow over bedrock.

Near Lijiang, the soils are rather highly leached red earths which are acid but with low base status and a lower organic matter content (1.5-2%). Weathering is deep on all but the steepest slopes, where mass wasting processes are active.

5. Permafrost

Permafrost is widespread at the higher elevations on the QTP. It takes three forms. Continuous permafrost is found above about 4700m, and this grades down-slope into discontinuous permafrost with abundant thaw lakes. From about 4000m a.m.s.l., isolated patches of permafrost may be found, surrounded by areas of seasonal frost extending down to about 4250m a.m.s.l. (Jin et al., 2009). The larger rivers and lakes are underlain by taliks (unfrozen ground surrounded by permafrost). About 1.75 x 10^6 km^2 of China is currently underlain by permafrost, of which 1.26 x 10^6 km^2 occurs on the Tibetan Plateau (Cheng et al, 2012; Li et al., 2012).

Figure 8. Collapsedpingo along the Golmud-Lhasa Highway in the Kunlun Pass.

The permafrost developed during the last two major glaciations, but has been undergoing periodic degradation since 11,800 years B. P. (Zhang et al., 1995; Pan and Chen, 1997; Jin et al., 2007). In general, the areas of alpine meadow correspond to those of discontinuous and sporadic permafrost, while the areas of continuous permafrost correlate with the areas of alpine semi-desert, desert and cushion plants. The permafrost profile may either include two layers (caused by a lower layer of relict permafrost) or a single layer more or less in equilibrium with the present climate. The former indicates present-day permafrost overlying a thawed zone, with relict permafrost beneath it. This is found primarily along the outer areas of continuous permafrost.

Associated with the permafrost are Greenland-type pingos, collapsed pingos (Figure 8), ice wedge casts, sand and rock tessellons (Cheng et al., 2006; Harris and Jin, 2012), ice wedge polygons, block streams (Harris et al., 1998), and cryoturbations (Cui, 1982). Gelifluction and cryoturbation are common, together with minor icings along the streams and adjacent to springs. On the northern slopes of the mountains above 5000m, cold-based glaciers occur.

There is an enormous body of literature on the changes in permafrost distribution on the QTP over the last 50 years (e.g., Cheng and Wu, 2007). A good summary of the evidence will be found in Jin et al. (2012). The aerial extent of permafrost on the eastern QTP has decreased from $1.5x10^6$ km^2 in 1975 to about $1.26x10^6 km^2$ in 2006. At Madoi on the eastern slopes of the QTP, the lower limit of permafrost has moved at least 15 km westwards upslope. Overall, the lower elevational limit of permafrost has risen by 40-80m. Along the Qinghai-Tibet Highway, the southern limit of island permafrost has moved 12 km to the north, while the northern limit has only moved 3km southwards. The mean annual ground temperature at the depth of zero amplitude has increased by 0.1-0.3˚C (Wang, 1993; Wang et al., 2000; Cheng and Wu, 2007), while the thickness of the active layer has increased by 10-40 cm (Wu and Liu, 2004; Yang et al., 2004; Wu et al., 2006). The mean annual ground temperature in the soil overlying permafrost increased from 0.1-0.7˚C from 1996 to 2001 (Wu and Liu, 2004; Wu et al., 2006).

Seasonal frost has also changed in areas outside the permafrost areas (Wang and Harris, 2012), based on observations started in 1961 at the weather stations using frost tubes. Since 1961, the maximum depth of frozen soil has decreased by 0.14-1.0m at 22 sites between 2500-4500m a.m.s.l., while the annual period with frozen soils decreased by about 35 days. Maximum depth of freezing increased towards the northwest, but the length of ground freezing was more variable, though showing a similar general trend. The changes in seasonal freezing appear to be due to a warming of winter minimum air temperatures, especially at the higher altitudes. The changes in length of ground freezing are greatest in the Fall and least in the Spring, probably due to the latent heat needed for melting snow and ice in Spring. This evidence also suggests a weakening of the effects of the cold Siberian cP/cA air mass over central China.

6. Hydrology

The hydrology of the upper reaches of the Yangtze River is a function of the climate, topography, glaciers and permafrost. The latter stores large quantities of water in the form of ice in the ground. Furthermore, the permafrost acts as an impermeable layer separating supra-permafrost ground water from the main water table below the permafrost. Seasonal frost affects the distribution of near-surface soil moisture, as well as the timing of thawing of the ground. Anthropological activities have also affected the hydrology. A good summary of the relationship of permafrost to groundwater on the Plateau and in North-east China will be found in Cheng et al. (2012).

The upper western reaches of the Yangtze River can be divided into two parts, viz., upper part of the river crossing the Plateau, and the part where the river descends the eastern slopes of the QTP to the Sichuan Basin at Yichin. On the Plateau, the river is a rather small stream flowing north, away from the Tanggula Range, before turning east to traverse the northeast central part of the Plateau. There are numerous isolated thaw lakes and water bodies occupying deflation hollows scattered across the landscape. The limited precipitation produces rather muddy conditions in the lower parts of the landscape due supra-permafrost water perched on the underlying permafrost table. Along the rivers and under the larger lakes, through taliks occur, some allowing hot and warm springs in the lower places where sub-permafrost waters can come to the surface under artesian pressure. East Greenland-type pingos occur along some water courses, generally above about 4500m a.m.s.l. where there is continuous permafrost with few through taliks (Figure 8). They tend to degrade as soon as the flow of spring water diminishes, so they are rather short-lived.

On the eastern slopes of the QTP, conditions change rather rapidly. River gradients increase, as does the frequency of precipitation from the East China Summer Monsoon. The semidesert grades into lush alpine *Kobresia* meadow lining the lower parts of the valleys at about 4200m a.m.s.l. (Figure 4). The slopes along the sides of the valleys increase, resulting in increased water erosion if the vegetation cover is damaged. The main streams in the area become deeply entrenched on the mountain sides, though at least 15 reservoirs have been built along the major streams and their tributaries (Jin et al., 2010). At lower elevations, the lush mixed coniferous forest occurs which represents one of the two main forested areas of China. Deforestation and agriculture have taken place on the accessible lower slopes. In the upper Jialing basin, 70% of the farmland is found on slopes of 20-30° (Dingzhong and Ying, 1996). Cultivated slopes may account for 50-90% of the total farmland, with 33% having slopes in excess of 25°. The heavy monsoon rains cause enormous erosion of the topsoil by rain splash and sheet wash due to the poor vegetation cover. Maximum erosion rates may reach 9,000-36,000 t km^2 a in some places.

Avalanches, landslides and debris slides and flows are common on the steep slopes and unstable banks. These are aided by thunderstorms and earthquakes in areas with easily erodible deposits. In the Upper Jialing basin, there are about 10,000 sites with landslides. Maximum volume of material moved may be as large as 2,000,000 m^3 of soil and rock which moved during the Xintan landslide on the banks of the Yangtze River on 12th June, 1985. Debris flows are abundant, with over 3,000 being recorded in the Upper

Jialing basin. They contribute about 6,000,000 t of sediment to the Jinsha River each year.

7. Pollution

The Yangtze River on the QTP is relatively unpolluted compared with the rest of the Yangtze River basin. The sporadic mining of sediments in search of gold has been outlawed, though some pollution may access the river from the settlements along the Qinghai-Tibet Transportation Corridor. On the east slopes of the QTP, there are only sparse settlements which do not seriously affect water quality. However, from Lijiang to Yichang, population pressures start to build and there is significant pollution with high concentrations of arsenic, cadmium, lead and zinc (Song et al., 2010). The mountains south of the river and east of the Sichuan Basin are rich in mineral deposits, and provide significant quantities of these and other heavy elements to the sediment being transported by the river and its tributaries.

8. Population

The lower sections of the mountain slopes and valleys near Lijiang are home to several ethnic minorities as well as the Han. One of the main arguments against the Hutioaxia (Tiger Leaping Gorge) Dam was the fact that it would displace 100,000 mainly Naxi minority farmers in a region where resettlement on the surrounding mountains would be difficult, if not impossible. The local population had seen what had happened to those displaced by the lake behind the Three Gorges Dam and objected rather effectively. Currently, the project is being moved 200 km upstream where it will displace 20,000 Tibetans. The latter are the main ethnic group from there westwards to the QTP. Downstream of Lijiang, the Han become the most numerous ethnic group. On the QTP, Tibetan herders are the dominant group, though Han are usually found in the settlements along the Qinghai-Tibet Transportation Corridor. The nomadic herders that successfully made a living moving around the QTP with their herds were forced to change their ways (see Agriculture below).

UTILIZATION OF THE REGION

There have been substantial changes in land use in the last 50 years. Urbanization has been limited in the region, but the development of transportation corridors has provided better access to the QTP from the north and east. However, the effects of climate change have forced the central administration to change the land use in order to minimize desertification. On the forested eastern margins of the QTP, changes in government priorities and their effects have necessitated remedial measures including reforestation, curtailing widespread plundering of the forest resources, and tighter controls on mining. Large protected areas have been established. These will be dealt with in the following section.

1. Nature Reserves

One of the most important is the Sanjiangynuan National Nature Reserve (SNNR) or the Three Rivers Nature Reserve in the Qinghai Province. It includes the headwaters of the Yellow River, the Yangtze River and the Mekong River on the QTP. It is a complex of 18 subzones, each with three zones managed with differing degrees of strictness. Within these areas, uncontrolled mining, logging, hunting and grazing have been curtailed, though not entirely eliminated. Foreign and other mining firms have replaced the uncontrolled miners, measures have been taken to protect endangered species, and trees have been planted where conditions are appropriate. Grazing of domestic herds is only permitted outside the core zones, and then only under strict supervision. Rangeland is being fenced and residents and nomads in the core zones are being resettled elsewhere (Banks et al., 2003; Foggin, 2008). Unfortunately, not all the Tibetan herders want to move into the new housing provided on grasslands in southern Gansu Province (Banks, 2003). Altogether some 200,000 people are being resettled.

A second one is the Three Parallel Rivers of Yunnan Protected Areas, a UNESCO World Heritage Site. It lies within the drainage basins of the Yangtze, Mekong and Salween Rivers in the Chinese portion of the Hengduan Mountains in Yunnan Province. It consists of 15 separate core areas with buffer zones. Total area protected is 939,441.4 ha (9394.4 km^2) with 758,977.8 ha (7589.78 km^2) of buffer zones. It is where the three great rivers run parallel to one another for about 300 km.

There are also about 39 Giant Panda Reserves in the more inaccessible locations in the mixed coniferous forest zone in Yunnan, Gansu, Sichuan and Shangxi Provinces. The two main ones are the Sichuan Giant Panda Reserve at Wolong and the Bifengxia Panda Base. The Wolong National Natural Reserve is located about 3 hours drive north of Chengdu. It began in 1963 and is the first (established in 1963), largest and best-known panda reserve and was placed on the UNESCO Man and Biosphere Reserve Network in 1980. Unfortunately the visitors are inadvertently damaging the panda habitat (Liu, 2001). Further hazards are frequent, including severe earthquakes. They occur throughout the region, and the one on May 12th, 2008 killed 5 security guards and one panda. This resulted in the captive pandas being temporarily relocated to the Bifengxia Panda Base.

2. Agriculture

The main agricultural area in the basin used to be the QTP. For centuries, the Tibetans lived a nomadic life in harmony with the environment. They were careful to move before a given area became too overgrazed and they maintained the stock levels such that it did not exceed the carrying capacity of the region (Scholz, 1995). In the eastern part of the QTP, Yaks are the preferred herd with smaller numbers of sheep and goats, plus a few horses. Westwards, the composition changed, with sheep and goats being more common in the drier, higher areas of alpine semi-desert (Miller, 1998). Since the sheep and goats crop the vegetation more closely, the question of the cause and effect of herding and semi-desert is a problematic question.

Recent studies (Klein et al., 2007) have shown that reasonable amounts of grazing are beneficial, improving rangeland quality and increasing net aboveground productivity (ANPP) by 20-40 g per m^2 per year without changing the percentage of palatable ANPP. It extends the growing season at both ends of summer and results in better nutrient quality (foliar nitrogen and carbon content). Thus the Tibetans were doing the right things to maximize animal production on the QTP. However the same research shows that experimental warming of the air decreases ANNP by 40 g per m^2 per year on the alpine meadows and decreased the percentage of palatable ANPP by 10 g per m^2 per year. Furthermore, there was a significant change in species composition from primarily palatable grass to unpalatable shrubs. Shrubs have only 25% digestible dry matter compared with 60% digestible dry matter in graminoids. The conclusions of this study are that moderate grazing can mitigate the negative effects of warming by climate change, and is an important tool in keeping shrubland at bay. Another study comparing rotational grazing, continuous grazing and no grazing on the biomass in the *Stipa breviflora* Griseb desert grassland between 2005 and 2007 showed that rotational grazing alleviated deterioration of the grassland while preserving more above-ground and below-ground biomass (Yan et al., 2012).

This was not known to the Chinese administration in 1952 when they enacted sweeping reforms on the QTP. The central Government appropriated both the land and livestock and reorganized the former Tibetan herders into communes that were based in specific places. Livestock production increased but the areas around the communes became severely eroded within two years. In 1958, the system was abandoned, but by then, it was too late. Severe desertification was occurring on the QTP. On the upper part of the eastern slopes above tree line, pastoral land was often tilled, but the farms often were soon abandoned. Unfortunately, the fields have not been restored to their previous state.

To try to overcome these problems, the Tibetan pastoralists were given back their flocks and herds, but not their land. The Tibetans promptly reduced the numbers of stock to more reasonable numbers and tried to reuse their traditional herding. However, much of the land had been degraded (Figures 6 and 9), and so the Government commenced a program of moving people "out of poverty" and resettling them in newly and hastily built villages in grasslands in Gansu Province and elsewhere. Substantial areas are currently not used for grazing.

A policy of fencing grazing lands was introduced, but the owners still grazed their herds and flocks on communal lands, using the fenced fields to produce fodder for the winter. The increasing air temperatures have made the situation more difficult, while the heavier winter snowfalls are resulting in increased winter livestock mortality. The bare patches of soil are subject to severe wind erosion and there has been a population explosion of pikas. Attempts have been made to poison the pikas but this is against the Tibetan religion. Ground cracking in the winter further aids erosion. These problems are discussed in numerous publications, e.g., Williams (1996), Miller (1998), Sneath (1998), and Ho (2000).

The changes in air temperature during the last few decades have altered the phenology of the forbs. Although the winter snows melt earlier, the forbs still do not change their time of growth significantly (Yu et al., 2010). This has prompted various theories by biologists as to why this occurs (Chen et al., 2011), but it is almost certainly due to increased accumulation of water in the form of ice in the soil during the winter. Lack of a vegetation cover increases soil cooling (Hu et al., 2008) and the increased winter snowfall provides more water. This water moves to the coldest places in the soil, and in spring, it has to be changed back from ice to water at 0°C, and this uses more solar energy before the ground can warm up sufficiently for the forbs to start growing.

3. Mining

During the Great Leap Forward (1958-1961), the Chinese people were encouraged to produce as much iron as possible. The lateritic soils south of the Lijiang area were one of the sources of iron, and considerable deforestation occurred both to find the ore and for small-scale smelting (Biot and Lu, 1993), the so-called "backyard furnaces". In subsequent five-year plans, the emphasis was in other economic areas, and these operations largely ceased, and this has been replaced by the mining of a variety of metaliferous ores. The valley of the Upper Yangtze east of Yibin and in the southern area southeast of the Tiger Leaping Gorge is very rich in minerals. These include copper, lead, zinc, stibium, tungsten, cobalt, tin, iron, manganese and aluminum.

Figure 9. Yaks grazing on eroded alpine steppe.

The area produces approximately 30-50% of the total Chinese output. In addition, it accounts for 80% of Chinese reserves of mercury, magnesium, vanadium and titanium (Wei and Jin, 2006). As a result, the forests in this area has been severely damaged by mining operations, resulting in greatly increased erosion of the landscape with a corresponding increase in suspended sediment load in the tributary streams entering the Yangtze River.

There is no significant mining on the QTP. A few transient gold seekers try to find placer gold near the Kunlun Shan but with limited success.

4. Dams

In order for China to continue to grow and double its gross domestic product (GDP) by 2020, it has to double its energy supply. In order to meet its goal, China has to bring on-stream the equivalent on Three Gorges Dam every two years. An obvious place to generate electricity is along the Upper Yangtze River as it descends from the QTP (Yao et al., 2006).

The first hydropower dam, the Shilong Dam was built on the River in 1910, and a further 31 small dams were built by 1949 (Wei and Jin, 2006). At least 15 dams have been constructed on the Yangtze River and its tributaries within the study area (Jin et al., 2010). There are part of a proposed cascading sequence of dams to produce hydroelectric power and control flooding, though some authors acknowledge that the western part of the upper reaches of the Yangtze River do not contribute appreciably to the flooding downstream. A list of some of the current dams under construction on the upper reaches of the Yangtze is provided on the International Rivers website.

As discussed by Yao et al., a major problem with these projects is the displacement of the residents in the substantial areas being flooded. There is low compensation for dam resettlement, and the new settlers tend to not be accepted by the existing residents of the resettlement area. The Yunnan Provincial Government has the power to evict the villagers whose ancestors have lived at that site for centuries, but there is no mechanism to protect these local residents when they move. Stojanov and Novosak (2006) give more details of the ensuing problems, but in an area with extremely steep slopes, resettlement in the adjacent area is not possible.

The proposed dam 200 km upstream of Tiger Leaping Gorge will only displace 200,000 people, but it will create a long lake, extending up-valley into the Three Parallel Rivers of Yunnan Protected site. Furthermore, cascading dams will have an enormous impact on the aquatic life in the Yangtze River and its tributaries (Wei and Jin, 2006). Since China is home to one of the most diverse faunas and floras in the world, developing the cascading dams without concern about the ecology of the river is likely to result in a substantial loss of biodiversity. A further concern is the fact that these dams are being built in an area that suffers frequent strong earthquakes. Presumably the theory of cascading dams is that those downstream of a catastrophic earthquake should be able to hold the sudden flow of water when a dam breaks. If not, then the settlements downstream will have more than the earthquake shocks to deal with.

5. Transfer of Water from One River Basin to Another

The Yellow River has been drying up in parts of its lower reaches during the early part of the 21st millennium, but is relied on for water along the eastern margin of the desert in Gansu Province. Accordingly, the Chinese Government has been studying the feasibility of transferring water from the headwaters of the Brahmaputra (Yalu Zangpo, or Yarlung Zangpo), Mekong (Lancang) and Yangtze Rivers to the headwaters of the Yellow River and Hai River. The idea originated with Mao Zedong, and the feasibility of this has been extensively studied 44.8 billion cubic meters of water a year is involved, moving along three different routes, of which this is the western one (the West Line).

The first part of the project would entail moving water through the drainage divide between the Yangtze and Yellow rivers in the headwater area of the latter. The water would have to be stored behind dams and then moved through a system of long tunnels and associated dams. Loss to evaporation would be considerable, but the north-east of industrialized China lacks sufficient water for its current population. It also needs water to combat desertification and for the growing demands of industry. It has been suggested that an additional 200 billion cubic meters of water a year could be diverted from the Mekong, Yarlung Zangbo and Salween Rivers into the system (Craig Simmonds, 2008).

This has been severely criticized by environmentalists (Craig Simmonds, 2006) and obviously would be viewed unfavorably by the neighboring countries to the south.

Figure 10. Farms along the lower slopes of the Yangtze River above the Leaping Tiger Gorge.

6. Forestry

The forests on the eastern slopes of the QTP are part of one of the two largest forested areas in China. The mixed coniferous forests consisted historically of dense natural forests and vast grasslands. After the founding of modern China in 1949, forest harvesting became the primary use of the areas that were reasonably accessible. Many forested areas were converted to farmlands where slopes were suitable (Figure 10). Timber harvests were particularly great between 1950 and 1970. Forest stock volume decreased 3.0 million cubic meters per year and forest coverage by 0.47% per year in the Minjiang watershed between 1949 and 1980 (Fan, 2002). As a result, the dominant vegetation types are meadows (36.6%), coniferous forests (30.7%), and shrublands (deforested areas – 22.8%). These make up 90% of the land cover (Zhang et al., 2008).

The major challenge in this region is to prevent the destruction of the forest since reforestation tends to be difficult. The logged areas were not replanted and natural regrowth of the forest has been found to be slow. Shrublands replace the forest unless tree seedlings are planted. Species appear from other regions as weeds, and the impacts of changes in vegetation cover on the hydrology can be as great as that of climate change (Cui et al. 2012). Reforestation reduces soil erosion by reducing runoff and conserving soil water.

Conclusion

The western section of the Upper Yangtze River basin is a critical part of China. It has the potential of providing much-needed, good-quality water to the arid north part of China, while also being a significant potentially large grazing area in a country with a burgeoning population and limited agricultural land. The enormous change in elevation between the QTP and the Sichuan basin provides enormous potential for the generation of electricity, even though it is rather a long way from the main industrial centers. This can only be maximized by significant displacement of minority groups although the current population density is low. Undoubtedly this area will be one that will see significant changes in land use during the next few decades, and these should also include reforestation of the eastern slopes of the QTP to provide a replenishment of the supply of timber in one of the two main forested areas of China. Being in the subtropics, the growth of the trees in the deforested area should be more rapid than in northeast China.

Acknowledgments

The author is indebted to an anonymous Chinese reviewer who kindly made suggestions to improve this Chapter. Dougliang Luo drew Figure 1, and the photographs are by the author.

References

Banks, T., 2003. Property Rights Reform in Rangeland China: Dilemmas on the road to Household Ranch. *World Development,* vol. 31(12): 2129-2142.

Banks, T., 2003. Community Based Grasslands Management in Western China. *Mountain Research and Development,* vol. 23(2): 132-140.

Breckle, S.-W. and Rafiqpoor, M. D., 2010.Field Guide Afghanistan: Flora and vegetation. Bonn, *Scientia Bonnensis*, 831p.

Chen, H., Zhu, Q., Wu, N., Wang, Y. F. and Peng, C. H., 2011. Delayed spring phenology on the Tibetan Plateau may also be attributable to other factors than winter and spring warming. *Proceedings of the National Academy of Sciences,* vol. 108, #9: E93.

Chen, S., Lui, Y. F. and Thomas, A., 2006. Climatic change on the Tibetan Plateau: Potential evapotranspiration trends from 1961-2000. *Climatic Change,* vol. 76: 291-319.

Chen, X., Zong, Y. Q., Zhang, E., Xu, J. G. and Li, S., 2001. Human impacts on the Changjiang (Yangtze) River basin, China, with special reference to the impacts on the dry season discharges into the sea *Geomorphology,* vol. 41: 111-123.

Cheng, G. and Jin, H., 2012. Permafrost and groundwater on the Qinghai-Tibet Plateau and in northeast China. *Hydrogeology Journal,* in press. DOI:10.1007/s10040-012-0927-2.

Cheng, G. and Wu, T., 2007. Responses of permafrost to climate change and their environmental significance, Qinghai-Tibet Plateau. *Journal of Geophysical Research,* 112:F02S03, 10pp. DOI:10.1029/2006JF000631.

Cheng, J., Zhang, X., Tian, M., Yu, W. and Yu, J., 2005. Ice-wedge casts showing climatic change since the Late Pleistocene in the source area of the Yellow River, north-east Tibet. *Journal of Mountain Science,* vol. 2(3): 193-201.

Craig Simmonds, 2006. In China, a water plan smacks of Mao. Cox News Service. http://webarchive.org/web/20070911233235/http://coxwashington.com/hp/content/reporters/stories/2006/09.

Craig Simmonds, 2008. *Solving the entire Chinese water crisis.* The Atlanta Journal-Constitution.

Cui, X., Liu, S. and Wei, X., 2012. Impacts of forest changes on hydrology: a case study of large watersheds in the upper reach of the Yangtze River Basin. *Hydrology and Earth Systems Sciences Discussions,* vol. 9: 6507-6531.

Cui, Z, 1982. Basic characteristics of periglacial landforms in the Qinghai-Xizang Plateau. *Scientia Sinica,* vol. 25(1): 79-95.

Ding, Y. J., Liu, S. Y., Xie, C. W., Zhang, Y. and Wang, X. D., 2006. *Yellow River at risk: An assessment of the impacts of climate change on the*

Yellow River source region. www.yellowriversource.org (For Greenpeace).

Dia D. and Tan, Y., 1996. Soil erosion and sediment yield in the Upper Yangtze River basin. In:*Erosion and sediment yield: Global and regional perspectives.* Proceedings of the Exeter Symposium. IHAS Publication #236: 191-203.

Fan, H., 2002. A study on 50 a land use and cover change of watershed of Upper Mijiang River. *Journal of Mountain Science,* vol. 20: 64-69.

Fang, X., Zhu, L., Wang, C., Zhao, X. X., Chen, H. and Yang, W., 2006. Quaternary sediment in the Yichang area: Implications for the formation of the Three Gorges of the Yangtze River. *Geomorphology,* doi: 10. 1016/ geomorph. 2006.03.027.

Foggin, J. M., 2008. Depopulating the Tibetan Grasslands. *Mountain Research and Development,* vol. 28(1): 26-31.

Harris, S. A., Cheng, G., Zhao, X, and Yongqin, D., 1998. Nature and dynamics of an active block stream, Kunlun Pass, Qinghai Province, People's Republic of China. *Geografiska Annaler,* vol. 80A(2): 123-133.

Harris, S. A. and Jin, H. J., 2012. Tessellons and "sand wedges" on the Qinghai-Tibet Plateau and their palaeoenvironmental implications. *Proceedings of the Tenth International Permafrost Conference, Salekhard,* vol. 1, pp. 149-153.

Ho, P., 2000. The clash over State and Collective Property: The making of Rangeland Law. *The China Quarterly,* #161: 240-263.

Hu, H., Wang, G., Liu, G., Taibing, R. D., Wang, Y, Cheng, H. and Wang, J. F., 2008. Influences of alpine ecosystem degradation on soil temperature in the freezing- thawing process on Qinghai-Tibet Plateau. *Environmental Geology,* vol. 57: 1391-1397.

Institute of Soil Science, 1986. *The Soil Atlas of China.* Institute of Soil Science, Academica Sinica. Beijing, China, Cartographic Publishing House. 42p.

Jin, H., Chang, X. and Wang, S., 2007. Evolution of permafrost on the Qinghai-Xizang (Tibet) Plateau since the end of the late Pleistocene. *Journal of Geophysical Research,* vol. 112, F02S09, doi: 10. 1029/ 2006J F000521.

Jin, H., He, R., Cheng, G., Wu, Q., Wang, S, Lin, L. and Chang, X., 2009. Changes in frozen ground in the source area of the Yellow River on the Qinghai-Tibet Plateau, China, and their eco-environmental impacts. *Environmental Research Letters,* vol. 4: 1-11.

Jin, H., Yu, Q., Wu, J. and Wang, S., 2011. Spatiotemporal variability of permafrost degradation on the Qinghai-Tibet Plateau under a warming climate. *Sciences in Cold and Arid Regions,* 3(4): 281-305.

Jin, Y., Zhou, J. J. and Huang, G. X., 2010. Impact of large reservoirs operation on the hydrological process in the upper Yangtze reaches. *Journal of Hydroelectric Engineering,* vol. 29(2): 94-101.

Kirby, E. and Ouimet, W., 2012. Tectonic geomorphology along the eastern margin of Tibet: Insights into the pattern and processes of active deformation adjacent to the Sichuan Basin. *Geological Society of London,* Special Publication vol. 353: 165-188.

Klein, J. A., Harte, J. and Zhao, X. Q., 2007. Experimental warming, not grazing, decreases rangeland quality on the Tibetan Plateau. *Ecological Applications,* 17(2): 541-557.

Li, H. and Wu, S., 1991. The floristic structure of the Xizang part of the Qinghai-Xizang Plateau plant subkingdom. *Chinese Geographical Science,* vol. 1(3): 221-233.

Li, L., Yang, S., Wang, Z., Zhu, X. and Tang, H., 2010. Evidence of warming and wetting climate over the Qinghai-Tibet Plateau. *Arctic, Antarctic and Alpine Research,* vol. 42(4): 449-457.

Li, X., Ran, Y., Nan. Z. and Jin, H., 2012. A frozen ground map of China made by integrating Existing permafrost maps and modeling results. In: Ma, W. and Niu, F. J., Eds. *Research Monograph of cold and arid regions engineering and environment: a collection Papers dedicated to the 70th Birthday of Academician Guodong Cheng.* Lanzhou, Lanzhou University Press, 229-239.

Liu, J., 2001. Endangered pandas not safe in Chinese Nature Reserve. *Environmental News Service,* April 6th, 2001.

Liu, X. and Hou, P., 1998. Relationship between the climatic warming over the Qinghai-Tibet Plateau and its surrounding areas in recent 30 years and elevation. *Plateau Meteorology,* vol. 17: 245-249.

Lϋ, X. and Lϋ, X., 2002. Climate Tendency Analysis of warming and drying in Grassland of Northeast Qinghai-Tibet Plateau of China. *Grassland China,* vol. 24: 8-13.

Miller, D., 1998.Tibetan Pastoralism: Hard times on the plateau. *Chinabrief,* vol. 1(2): 17-22.

Pan, B. and Chen, F., 1997. Permafrost evolution in the Northeastern Qinghai-Tibet Plateau during the last 150,000 years. *Journal of Glaciology and Geocryology,* vol. 19(2): 124-132. [In Chinese with an English abstract].

Plateau Perspectives, no date. *http://www.plateauperspectives.org/.*

Qinghai Geological Survey and China University of Geology, 2002. *Investigation Report on the Ecological Environmental Geology in the Sources of the Yellow River.* Xi'ning, Qinghai Geological Survey, China, 50p. [In Chinese].

Ren, S., 1996. *Several major achievements in early Neolithic China, ca. 5,000 BC.* http://http-server.carlton.ca/~bgordon/Rice/papers/REN96.htm.

Richardson, N. J., Densmore, A. L., Seward, D., Wipf, M. and Yong, Li, 2010. Did incision of the Three Gorges begin in the Eocene? *Geology,* vol. 38 (6): 551-554.

Schulz, F., 1995. *Nomadismus. Theorie und Wandel einer sozio-ökologischen Kulturweise.* Stuttgart, Germany, Franz Steiner Verlag.300p. [In German].

Sneath, D., 1998. State Policy and pasture degradation in Inner Asia. *Science,* vol. 281: 1147-1148.

Song, Y., Ji, J., Mao, C., Yang, Z., Yuan, X., Ayoko, G. A. and Frost, R. L., 2010. Heavy metal contamination in suspended solids of Changjiang River – environmental implications. *Geoderma,* vol. 159: 286-295.

Stojanov, R. andNovosák, J., 2006. Environmental migration in China. *Geographica,* vol. 39:65-82.

Tan, C., Liu, X. and Wang, N., 2000. Climate change range of Qinghai-Tibet Plateau. *Chinese Science Bulletin,* vol. 45: 98-106.

Varis, O., Kummu, M. and Salmivaara, A., 2012. Ten major rivers in monsoon Asia-Pacific: An assessment of vulnerability. *Applied Geography,* vol. 32: 441-454.

Wang, K., Sun, J., Cheng, G. and Jiang, K., 2012. On the influence of altitude and latitude on mean surface air temperature across the Qinghai-Tibet Plateau. *Journal of Mountain Science*, vol. 8: 808-816.

Wang, S., 1993. Recent change of permafrost along Qinghai-Xizang Highway. *Arid Land Geography,* vol. 16: 1-7.

Wang, S., Jin, H., Li, S. and Zhao, L., 2000. Permafrost degradation on the Qinghai-Tibet Plateau and its environmental impacts. *Permafrost and Periglacial Processes,* vol. 11: 43-53.

Wei, H. and Jin, C., 2006. A preliminary approach on impacts of hydropower development in Upper Yangtze River on ecological environment. *Hydropower,* 2006: 1098-1106.

Willams, D. M., 1996. The barbed walls of China: A contemporary grassland drama. *The Journal of Asian Studies,* vol. 55(3): 665-691.

Wilson, C. J. L. and Fowler, A. P., 2011. Denudation response to surface uplift in east Tibet: Evidence from apatite fission-track thermochronology. *Geological Society of America Bulletin,* vol. 123(9-10): 1966-1987.

Wu, Z, 1985-1987. *Flora Xizangica.* Beijing. Science Press, 5 volumes [In Chinese].

Wu, Q. and Li, W., 2004. Ground temperature monitoring and its recent change in Qinghai-Tibet Plateau. *Cold Regions Science and Technology,* vol. 38: 85-92.

Wu, Q., Cheng, G., Ma, W., Niu, F and Sun, Z., 2006. Technical approaches on permafrost stability for Qinghai-Tibet Railway. *Geomechanics and Geoengineering, An International Journal,* vol. 1(2): 119-127.

Xin, X., Zhou, T. J. and Yu, R., 2010. Increased Tibetan Plateau snow depth: An indicator of the connection between enhanced winter NAO and late-Spring Tropospheric cooling over East Asia. *Advances in Atmospheric Sciences,* vol. 27(4):788-794.

Xu, C., Liu, J., Song, L., Jiang, W. and Shi, C., 2000. GPS measurements of present-day uplift in the Southern Tibet. *Earth Planets Space*, 52: 735-739.

Yan, R., Xin, X., Wei, Z., Liu, S., Yang, J. and Yang, G., 2012. Influences of various grazing systems on community biomass of a desert grassland in Inner Mongolia. *Sciences in Cold and Arid Regions,* vol. 4(4): 303-309.

Yang, J., Ding, Y., Chen, R. and Shen, Y., 2004. Permafrost change and its effect on eco-environment in the source regions of the Yangtze and Yellow Rivers. *Journal of Mountain Science,* vol. 22(3): 278-285.

Yao, Y., Zhang, B., Ma, X. D. and Ma, P., 2006.Large-scale hydroelectric projects and Mountain Development on the Upper Yangtze River. *Mountain Research and Develop-ment,* vol. 26(2): 109-114.

Zhang, Y., Liu, S., Wei., X., Liu, J. and Zhang., G., 2008. Potential impact of afforestation on water yield in the subalpine region of Southwestern China. *Journal of the American Water Resources Association,* vol. 44: 1144-1153.

Zhang, Y., Zhang, J., Xu, J. and Lin, F., 1995. Paleoclimatic evolution of the Yellow River source areas since the Holocene Epoch. *Earth Science – Journal of China University of Geosciences,* vol. 20(4): 467-469. [In Chinese with an English abstract].

In: Yangtze River
Editors: T. B. Maloney, B. R. Hutchins
ISBN: 978-1-62618-285-1

Chapter 3

The Yangtze River: Hydro Development, Changing Geography, Cultural and Environmental Implications

Kepa Morgan* and Amaria Francis Waretini
The University of Auckland, Auckland, New Zealand

Abstract

The Yangtze River system is Asia's longest river and has its glacial source in Qinhai province on the Tibetan Plateau. The Yangtze flows eastward from its source 6,418 kilometres across the People's Republic of China (China) before discharging into the East China Sea. The Yangtze catchment is one-fifth of the total land area of China, making the Yangtze River one of the largest in the world in terms of discharge volume.

The Yangtze and over 700 tributaries continue to be the primary transportation link for one third of China's population (excluding the Grand Canal connectivity), and the main link for international shipping to the interior provinces of China. The rivers of primary importance to millions of people (more than thirty cities are located along the river) living on adjacent lands including the flood plains downstream of the Three Gorges Project. The river is administered by the Yangtze Forum which brings together 13 riparian provincial governments managing the

* Dr Kepa Morgan: Tel: ++ 64-9-3737999, email: k.morgan@auckland.ac.nz.

river from its headwaters to the East China Sea. Many cities extract and treat its waters for potable use, and improved riparian management encompassing reconstituting and reconnecting the myriad of disconnected lakes, watercourses and wetlands has protected the drinking water sources for several cities and also enhanced ecosystem habitat for many threatened species.

Despite improvements in the decision making processes, the Yangtze River remains the focus of on-going tensions regarding the sustainability and the appropriateness of the Three Gorges Project development completed in 2011. The Yangtze River system has been irreversibly altered from the “Golden Waterway” to a strategic asset now producing one-sixth of China's total electrical generating capacity supporting more than 20% of China’s GDP. The Three Gorges Project is a critical part of the plan China is implementing to increase its contribution to world trade and the standard of living for its people. River flow has been extensively modified with resulting implications in terms of reduced flood risks and sediment loads downstream, while the significantly raised water levels in the dam reservoir have forced the relocation of more than 1.2 million people and impacted upon geological stability.

The combination of historic relevance, impacted anthropological sites, ecosystem modification, altered river flows, industrial and urban uses, power generation, transportation reliability, geological instability, siltation issues, and economic base, under the Western World’s spotlight, are some of the inter-related attributes that make the Yangtze River a complex management challenge: A contemporary interpretation perhaps of its metaphor; the Golden Waterway.

The Mauri Model Decision Making Framework offers a new approach to the analysis of water resource management as it provides a transparent and inclusive approach to considering the environmental, economic, social and cultural aspects of decisions that have been taken regarding the Yangtze River system. The Mauri Model Decision Making Framework is capable of understanding multiple-worldviews and adopts mauri (intrinsic value, well-being or Qi) in the place of monetary equivalents as the base metric allowing an absolute sustainability assessment.

There are interesting parallels between the Yangtze River and the Waikato River in New Zealand, as while the actual physical scales are demonstrably different, the contextual relevance of the rivers resonates, as does the strategic importance of these rivers to their respective governments. The changing perceptions of both rivers from different worldviews, offers new ways of understanding sustainability, and better understanding the cultural and environmental implications of the development taking place on the Yangtze River.

A sustainability assessment for the Yangtze River focusing on the contemporary changes associated with the period of the Three Gorges

Project is presented and future opportunities are identified. The application of the Mauri Model Decision Making Framework to the Yangtze River reveals the parallels between the concepts of mauri and Qi, and the potential opportunities for the application of the framework in other Asian and Pacific contexts.

Keywords: Yangtze, Three Gorges Project, Water management, Environmental and Cultural Sustainability, Decision Making, Systems Thinking, Indigenous Knowledge, Mauri, Chi, Qi

INTRODUCTION

The Yangtze River system is Asia's longest river and has its glacial source in Qinhai province on the Tibetan Plateau. The Yangtze flows eastward from its source 6,418 kilometres across the People's Republic of China (China) before discharging into the East China Sea. The Yangtze catchment is one-fifth of the total land area of China, making the Yangtze River one of the largest in the world in terms of discharge volume. The Yangtze and over 700 tributaries continue to be the primary transportation link for one third of China's population (excluding the Grand Canal connectivity), and the main link for international shipping to the interior provinces of China. The river is of primary importance to millions of people (more than thirty cities are located along the river) living on adjacent lands including the flood plains downstream of the Three Gorges Project.

The Three Gorges Project is located on the Yangtze in the Hubei Province of China, and has been touted as the largest Civil Engineering project in the modern era (Saracino, 1997). The still functioning Dujiang Dam further upstream predates the Birth of Christ and was constructed for flood control and to provide irrigation (Jackson and Sleigh, 2001). The Dujiangyan irrigation system is noted for its innovative engineering approaches and layout that effectively excludes sediment (Cao et. al., 2010).China thus has an extensive historic association with hydraulic technologies for the goals of flood control and irrigation as well as facilitating navigation and water conservation (Needham, 1981). The Three Gorges Project (TGP) was first conceptualised to improve navigation and generate power in 1919 by Dr Sun YatSen, 'the father of modern China'. In 1932 a two month field survey in the Three Gorges area assessed the potential hydropower capacity in the Yangtze River.

In October 1949 the People's Republic of China formed, and just four months later the Yangtze Water Conservancy Committee was established in Wuhan, Hubei Province. Perhaps indicating mixed feelings regarding the implementation of an undertaking of such scale and complexity, Chairman Mao penned the following poem in 1956 (Jin, 2009);

> Walls of stones will stand upstream to the west,
> To hold back Wushan's clouds and rain.
> Until a smooth lake rises in the narrow gorges.
> The mountain goddess if she is still there,
> Will marvel at a world so changed.

Construction for TGP began in 1994, power generation commenced in 2003 following the completion of phase II, and the project was nearing completion in 2010. TGP has a designed flood control capacity of 22 billion m^3. TGP comprises 26 hydro turbine 700MW power generators, producing 85 billion kilo Watt-hours of electricity annually (Chai, 2007). TGP measures 2.3 kilometres in length and required more than ten million tonnes of concrete, and just as many tonnes of ice cubes used to reduce the excessive heat of hydration temperatures resulting from the combination of high ambient temperatures and large volume concrete pours during construction.

The design and planning for TGP took place at a time of increasing emphasis on limiting or mitigating the environmental impacts of development in China. Institutional change, in 1988, restructured the purpose of environmental protection at the central government level to better respond to environmental concerns, and in 1989 following a ten year trial, new Environmental Protection Laws were introduced. The World Bank, the Asian Development Bank and the UNDP all became more active in China's development, and encouraged the transfer of assessment methods and techniques from overseas (Wang et. al., 2003). Cost-Benefit Analysis (CBA) was adopted to evaluate the monetary value of environmental protection measures and economic benefits, in order to coordinate economic development and environmental protection (Qu, 1987). Thus China could be considered to have been as concerned about achieving sustainable development in its broader meaning as any other Nation at that time.

Furthermore, although assessment methods and techniques were being introduced from overseas, China had already incorporated what translates as the 'Three Simultaneities' (3Ss) concept into all construction projects more than a decade earlier (Wang et. al., 2003). The 3Ss concept is compulsory for all construction projects and central to the Chinese environmental protection system. The intention of 3Ss is to integrate the efforts to limit and mitigate potential pollution associated with a project's implementation across the three phases of designing, constructing and operating a project.

An analysis into the feasibility of TGP was completed by the CIPM Yangtze Joint Venture in 1988. TGP was designed with the goals of controlling flooding, improving navigation, generating electricity, and potentially providing water storage for irrigation (Jackson and Sleigh, 2001). This detailed study was based on an extensive Cost-Benefit Analysis, and determined that the project was technically feasible and should proceed (CIPM 1988: Vol. 1, 2-1).

Even prior to construction commencing in 1994, TGP was the focus of criticism in China and around the world. Could this be due to the limitations of CBA? Wang et. al (2003) point out that even though CBA has been required as part of the evaluation processes in China, the lack of prescriptive guidance on how to use this tool, encourages assessments that distort the analysis of benefits versus costs in favour of the more easily measured economic returns of the project. The assertion is that difficult to quantify, or unknown costs are omitted in assessments due to the uncertainty as to their likelihood or scale. For example in the case of TGP, the sedimentation of the dam reservoir, a problem that has forced the imminent closure of the ship transfer facilities for more than a year was difficult to predict and quantify accurately, or in more intrinsic terms, Chairman Mao's pondering whether the mountain goddess still remains.

The impacts of TGP are evidently very complex in nature and present as a complicated mix of issues and challenges, positive and negative, and to varying degrees. Despite improvements in the decision making processes, the Yangtze River remains the focus of on-going tensions regarding the sustainability and the appropriateness of the Three Gorges Project development completed in 2011. As a result the Yangtze River system has been irreversibly altered from the revered "Golden Waterway" of old, into a strategic asset now producing one-sixth of China's total electrical generating capacity supporting more than 20% of China's GDP. TGP is a critical part of the industrial plan that China is implementing to increase its contribution to world trade and the standard of living for its people. River flow has been extensively modified with resulting implications in terms of reduced flood risks and sediment loads downstream, while the significantly raised water levels in the dam reservoir have forced the relocation of more than 1.5 million people and impacted upon geological stability. Neither are these challenges were lost on the decision makers involved in TGP. In 2007, Deputy Director of the Construction Committee, Wang Xiaofeng, stated 'we absolutely cannot relax our guard against ecological and environmental security problems sparked by the Three Gorges Project' (Bristow 2007, and McCabe 2007). It is this environmental challenge and the extremely difficult to quantify cultural impacts that provide the impetus for revisiting the CYJV recommendation to proceed, although not limited to these aspects of sustainability in isolation.

With TGP substantially complete, the Yangtze River is now administered by the Yangtze Forum which brings together 13 riparian provincial governments managing the river from its headwaters to the East China Sea. Many cities extract and treat its waters for potable use, and improved riparian management encompassing reconstituting and reconnecting the myriad of disconnected lakes, watercourses and wetlands has protected the drinking water sources for several cities and also enhanced ecosystem habitat for many threatened species.

The combination of historic relevance, impacted anthropological sites, ecosystem modification, altered river flows, industrial and urban uses, power generation, transportation reliability, geological instability and siltation issues, all under the western world's spotlight, are some of the inter-related attributes that make the Yangtze River a complex management challenge: A contemporary interpretation perhaps of its metaphor; the Golden Waterway.

ASSESSING SUSTAINABILITY FOR THE THREE GORGES PROJECT

Sustainability can be defined as 'ensuring the needs of current generations are met without compromising the needs of future generations' (WCED, 1987). Intra-generational equity has become very important in project planning to ensure future generations are not hampered by poor decision making focused on short-term returns. Money is not an ideal metric for ensuring intra-generational equity. Money can not be used to assess long-term decision outcomes as its value diminishes over time, and measurements that rely on it, struggle to retain their reliability over time periods greater than five years.

Wang et. al. (2003) are likewise critical of sustainability decision making in China although for different reasons. They cite problems with Environmental Impact Assessment (CBA based) which 'limit its effectiveness as a tool for protecting the wider environment'. The narrow emphasis on pollution control to achieve environmental protection 'limits the attention given to other aspects of the biophysical and socio-cultural environment'. It is suggested that decisions are directed by technocrats, excluding local communities from involvement with the process. The lack of transparency is challenged as 'a recipe for the development of biased assessments and poor decision-making', especially when enforcement by national and local agencies is lacking. Thus a need exists for a simple, transparent, and effective sustainability assessment framework that can empower the voices of numerous stakeholders in decision making processes.

Sustainability assessment requires a holistic approach in order to effectively further the social, economic, environmental, and cultural spheres of well-being. The Mauri Model Decision Making Framework (MMDMF) was created to address the need for a decision support tool (DST) that met the needs of the Aotearoa New Zealand decision making context (see table 1). Further a good DST should enhance the way people perceive and understand a complex situation. The changed perception and understanding occurs primarily by modifying prior mental models as people's beliefs determine their actions.

Table 1. Essential Attributes for a DST in the Aotearoa New Zealand Context (Morgan, 2008)

Characteristic	Description
inclusive	effectively incorporate and represent Māori perspectives
indigenous	adopt a sustainability measure from indigenous thinking
eco-centric	demonstrate ecological integrity
holistic	acknowledge the interdependence of all life
equitable	deliver intra- and inter-generational equity
legal relevance	be effects focused promote social, economic, environmental, and cultural well-being
integrated	demonstrate interconnectedness between the criteria chosen
user friendly	be flexible yet easy to understand in its application
definitive	clearly determine whether a practice is or is not sustainable
transparent	be transparent and clearly identify applied bias

The attributes identified in the above table as essential for relevance in the Aotearoa New Zealand context, are also largely relevant in the assessment of TGP. The Mauri Model was conceived to meet these requirements and has been used in the previous chapter on the Waikato River to demonstrate its advantages in a complex cultural context.

A Mauri Model based analysis is an absolute sustainability analysis based on the impacts of a project on indicator 'Mauri'. The term 'Mauri' is analogous to the 'life force' within living things and the capacity to support life in air, water and soil. In the application of this model to the Chinese context, similarities have been observed between this concept of 'Mauri' and the Chinese concept of Qi. Qi is referred to as a word meaning 'breath', which is the 'cosmogenic stuff that gives birth to yin-yang', and is also the 'generative force of life and the world', according to Michael (2005, 18-20). It is further observed that there are many parallel concepts within the Māori and Chinese cultures (Morgan 2009), and hence the MMDMF can have cultural relevance in a Chinese context.

The MMDMF evaluates a project based on the impact on Mauri in four dimensions equivalent to economic, cultural, social and environmental well-being (Morgan 2006). The MMDMF is able to indicate result sensitivity to different worldviews or 'bias' by applying priority weighting to specific mauri dimensions determined using AHP pair-wise comparison, the Analytical Hierarchy Process (Saaty, 1980). AHP is a procedure that uses pair-wise comparisons to determine the relative importance of the four dimensions for each world view analysed. Rankings are determined using relative importance scores shown in Figure 1. If one dimension is extremely more important than another dimension, it would receive a '+3', while a dimension which is moderately less important than another dimension would receive a '-1'.

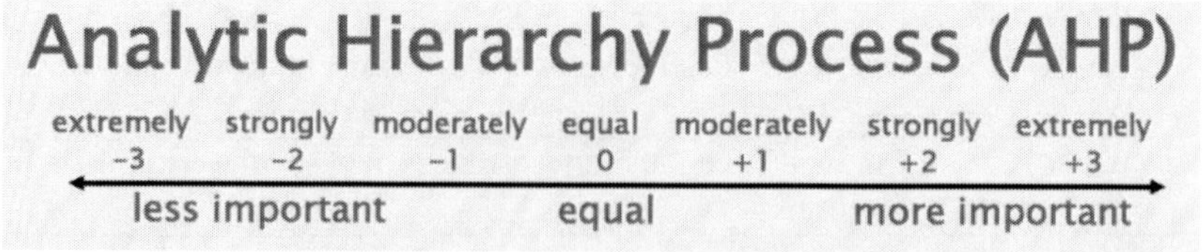

Figure 1. Relative Importance Scores for AHP Ranking of Stakeholder Group Perspectives.

Pair-wise Dimension Comparison	Ecosystem Qi (Environmental Well-being)	Village Qi (Cultural Well-being)	Community (Social Well-being)	Family Qi (Economic Well-being)	$\sum_{1}^{3} X + 9$	$\frac{W}{36}$
Ecosystem Qi (Environmental Well-being)	0	X_1	X_2	X_3	W	%
Village Qi (Cultural Well-being)		0				
Community (Social Well-being)			0			
Family Qi (Economic Well-being)				0		

Figure 2. Analytical Hierarchy Process (AHP) for Determining Stakeholder Priorities.

Table 2. Indicators Selected to Represent Three Gorges Project Impact Breadth

Qi Dimension	Dimension Indicators	Qi-ometer Rating
Ecosystem Qi (Environmental Well-being)	Physical Environment – Water Quality	
	Living Environment – Flora and Fauna	
	Environmental Geology and Erosion	
	Power Generation Displacing Coal	
Village Qi (Cultural Well-being)	Protection of Cultural and Historical Relics	
	Way of Life and Relevance of Knowledge	
	Identity - Changes to Regional Identity	
	Forced Relocation of Villages	
Community Qi (Social Well-being)	Quality of Life – Public Health	
	Regional Population Effects	
	Satisfaction/Strain and Happiness	
	Flood Reduction and Control	
Family Qi (Economic Well-being)	Economic Benefit/Cost of Project	
	Enabling Regional or National Benefit	
	Improved Navigation and Security	
	Individual or Family Development	

From the pair-wise comparison (Figure 2), the priorities of a variety of stakeholder groups can be quantified in terms of the mauri dimensions. The weightings then reflect how important each dimension is to each stakeholder group. For the purpose of this study, AHP was conducted for the viewpoints of the Chinese Government, Local Populations, and other Chinese People (outside of the region), in addition to a Western Perspective.

These are the main stakeholder perspectives that are relevant in terms of evaluating TGP and its impacts and the extant literature on TGP sustainability. The determined priorities for each group were based on background research, integrating perspectives expressed in the literature, and interviews with Chinese people regarding the importance of each of the four dimensions from their own personal viewpoints. In order to evaluate the sustainability of TGP, indicators are chosen that broadly represent each of the four dimensions. In this application of the framework, the Huhn Chinese concept Qi is adopted as the metric in place of mauri.

Raw ratings are determined by considering the impact upon the Qi of the indicators for each of the four dimensions. For TGP, there were a wide range of impacts, thus the indicators listed in Table 2 were intentionally selected to be as broad as possible, to ensure that all impacts were able to be assessed within the MMDMF.

The indicators are then scored with a rating of between +2 ('restored') and -2 ('destroyed') according to the Qi-ometer and decision tree provided in Figure 3. Determination of the rating follows the sequence shown and these ratings represent the impact (TGP has had) on those indicators.

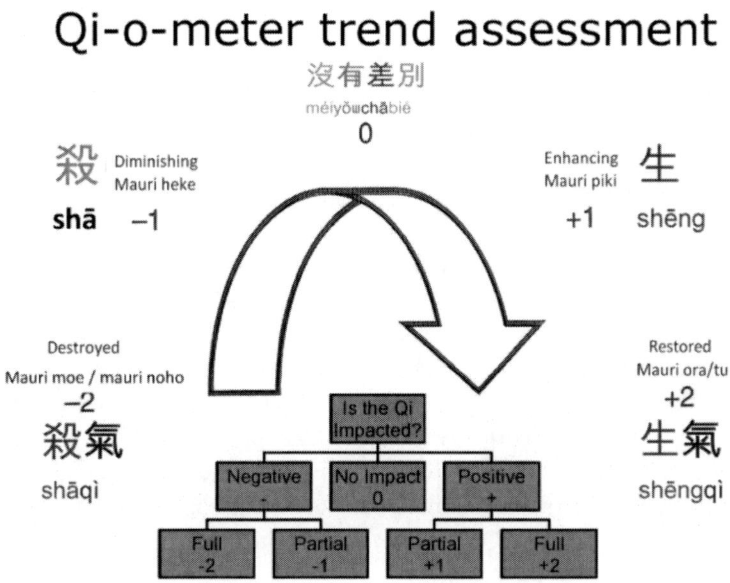

Figure 3. Qi-ometer for Evaluation of Absolute Impact on Indicator Qi.

Having determined the Raw Ratings for each indicator, the impacts for each dimension are determined as averaged values. These averages can be combined to reflect the overall holistic analysis for the physical and temporal situation being considered. These results are the actual unbiased and absolute understandings of the relative sustainability of each situation. While identified as raw rating averages, these results are in fact the objective understanding of the situation, albeit accepting the subjectivity of structuring reality as four dimensions of mauri or qi. Subsequently these dimension averages can be combined with the AHP results for each stakeholder bias to perform a sensitivity analysis on the results. The dimension averages are adjusted using the weightings previously determined in the AHP such that the level of perceived sustainability is quantified for each stakeholder perspective.

The MMDMF analysis therefore presents a Qi-ometer value for sustainability, based on the impact on Qi of the project. The significance of this result, and the influence of different stakeholder worldview priorities, can also be presented on the Qi-ometer.

THREE GORGES PROJECT ANALYSIS

The Qi-ometer shown in Figure 3 is a visual aide, demonstrating the absolute sustainability of a project. A sustainable project is considered to a score above zero for sustainability. The margin by which the result is greater than zero, indicates how strongly the project is enhancing or restoring the Qi of the indicators assessed. It would be unadvisable to proceed with a project whose assessment suggested that overall Qi or sustainability was being diminished or destroyed. Given the coarse nature of the Qi-ometer measurements, it is also wise to incorporate a reasonable margin above the zero result to provide a buffer, given that the some of the predicted outcomes can change over time due to incomplete information at the time a decision is taken. In the short term, it is also likely that the Qi-ometer assessment reflects an interim state that can not be ignored, for example during construction of the project. The changing quantum of the overall Qi-ometer result requires an approach that considers the cumulative impact on Qi. The approach adopted in this study is to repeat the Qi-ometer assessment of indicators for significant events in the history of the project.

With TGP, there are inevitable negative impacts during construction, before the anticipated benefits of the project are experienced. Some of the anticipated benefits are relatively immediate while others require periods of time to elapse to be experienced as improvements in Qi.

Thus the indicators are assessed for the following events:

1956: State of Qi at the time Chairman Mao penned his famous poem.

1992: State of Qi prior to Premier Li Peng handing final Three Gorges Project proposal to National People's Congress, using information available at the time the decision is made.

1993: Construction commences, impacts on Qi using information available in 1992.

2003: Change to Qi in 2003 when first power generation commences and reservoir begins to fill.

2012: The impact on Qi by 2012 of TGP, incorporating all information available since 1993.

2112: A '100 year' analysis, predicting the likely impacts on the Qi of each dimension 100 years after construction is complete.

The one hundred year forward prediction is chosen because although the period of evaluation for the CYJV Feasibility Study was limited to 50 years, the 100 year time period is better suited to the project type but ignores the flood return period of 1000 years for which TGP has been designed.

An explanation of the Ecosystem indicators is now provided. The *Physical Environment* indicator relates to pollution, water quality, sedimentation and climate change which are all physical impacts of the dam. The *Living Environment* includes wildlife and plants of all forms. The Chinese Alligator, the Finless Porpoise, the Chinese Paddlefish, and the Yangtze River Dolphin (Baiji) are all critically endangered species. The Baiji was declared functionally extinct after an extensive search of the river in 2006 revealed no signs of the dolphin since the construction of the dam, and many other wildlife and plants have been impacted. *Environmental Geology* has been affected with aggravated slope instability and erosion in the Hubei Province identified by geologists as an impact of the dam. Satellite images show significant changes to the surrounding topography since the construction of TGP. Hydro-electric *Power Generation* displaces the need for coal-fired power stations significantly reducing greenhouse gas emissions from the east, middle and south China with an 1000 kilometre effective supply radius covering almost 50% of the country (Jin, 2009).

The Village Qi indicators are now discussed. These are the indicators alluded to in Chairman Mao's poem and include; *Cultural and Historical Relics* including the impacts upon artefacts, temples or structures of historical significance e.g. tombs, archaeological sites. The most publicised of these are the Hanging Coffins, an ancient funeral custom where deceased relations are placed in coffins and hung from cliffs (Qing, 1998). Many of these sites were destroyed or inundated when the water level in the dam reservoir increased, however some have been salvaged and are now placed on display in museums associated with tourism. The indicator *Way of Life* and relevance of traditional knowledge represents changes in how people lived, and now no longer using traditional approaches of farming, trade, or travel. In particular the recurrence of floods and the mixed blessings that these brought along with the associated local wisdom are now of little relevance yet were intrinsic to survival previously.

Regional Identity speaks to the impacts and issues surrounding the foregone sense of belonging associated with long-term village communities now amalgamated as large urban centres with millions of people. The inherent loss of folklore, and changed identification of roles within traditional society, as the impacts of rapid urbanisation become apparent with future generations. Heritage and ancestral bond is considered as is the impact of rapid industrialisation which has irreversibly altered the roles of the elderly.

The *Forced Relocation of Villages* was necessary for approximately 1.5 million people and resulted in the disruption of connection to land for those being relocated and also those in areas identified for new settlement (Qigang, 1998). Many families involved in farming for centuries have been relocated to cities where their traditional source of purpose, livelihood and trade are now of little use. The social wellbeing indicator *Quality of Life* covers aspects such as public health and safety, access to health care and basic education. *Regional Population Effects* consider aspects such as the immigration of people and effects to achieve the construction challenge and resulting changes in how people live. *Satisfaction/Strain* addresses whether people experience greater happiness in their lives. Severe floods on the Yangtze River have been a regular recurrence approximately every six years in the past century. While the recorded loss of life has been reduced over time from the hundreds of thousands (1930's), to the tens of thousands (1950's), to less than 2000 people in 1998, the strain on those effected by loss of crops and infrastructure would have continued. *Flood Reduction and Control* acknowledges the reduced flood risk and the avoidance of significant future loss of life in downstream provinces protecting 15 million people and 2.3 million Hectares of farmland.

Economic indicators include the *Economic Benefit/Cost of Project* which addresses whether the project will generate economic returns for China. *Enabling Regional or National Benefit* acknowledges the indirect economic gains created due to enablers such as power, irrigation, and infrastructure secure from flooding.

Improved Navigation and security of supply encourages efficiencies in the movements of products and greater industrialisation. *Individual or Family Development* incorporates opportunities such as paid employment, higher skill levels and incomes, and that educated people will have greater earning potential. The assessment of indicator Qi impact is summarised in Table 3.

Table 3. Three Gorges Project Indicator Raw Ratings for Significant Events 1956 – 2112

Dimension	Indicators	1956	1992	1993	2003	2012	2112
Ecosystem Qi	Physical Environment	+2	+1	-1	-1	+1	+1
	Living Environment	+1	-1	-1	-2	-2	+1
	Environmental Geology	+1	+1	0	-2	-1	0
	Power Generation	-2	-2	-2	+1	+2	+2
Village Qi	Historical Artefacts	+2	+2	+1	-1	-2	-1
	Way of Life	+2	+1	-2	-2	-2	-1
	Regional Identity	+2	+2	0	-1	-2	0
	Location Permanence	+2	+2	-1	-2	-1	-1
Community Qi	Quality of Life	+1	+1	0	-1	+1	+2
	Population Effects	0	0	-1	-1	-1	+1
	Satisfaction/Strain	+1	+1	+1	-1	+1	+2
	Flood Exposure	-2	-1	-1	+1	+2	+2
Family Qi	Project Economics	-1	-1	-2	-2	+1	+2
	National Benefit	0	-1	-1	-1	+2	+2
	Improved Navigation	-1	-1	-1	+1	+2	+2
	Individual Opportunity	0	0	+1	+1	+2	+1
Overall Qi	Average of All Indicators	+0.50	+0.25	-0.63	-0.81	+0.19	+0.94

While the individual indicator results can be rationalised and understood as presented, the overall Qi trend depicted in Figure 4 is an over-simplification that does not fully reflect the complexity of the Three Gorges Project. It is useful to identify the trends for dimension averages as these contribute to the overall Qi trend and help to illustrate which dimensions are most strongly influencing changes in Qi over time.

The dimension averages are provided in Table 4 and represented graphically in Figures 5, 6, 7, and 8. The separated Qi dimension trends shown in figures 5, 6, 7, and 8 provide an indication of how Qi is most significantly affected at different stages of TGP.

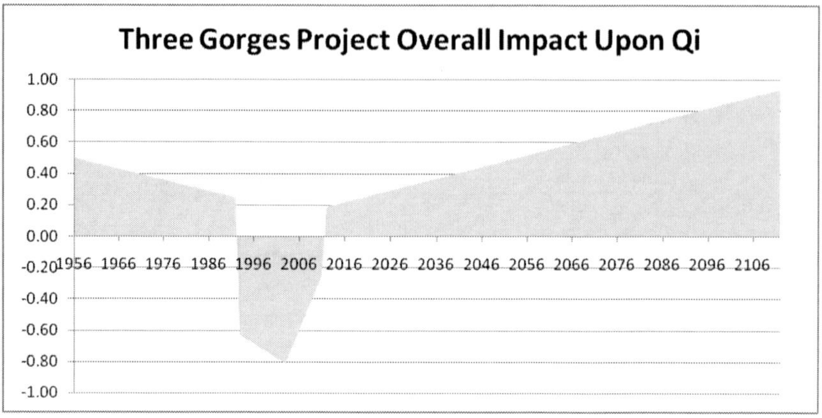

Figure 4. Overall Impact on Qi Determined for Three Gorges Project Using MMDMF.

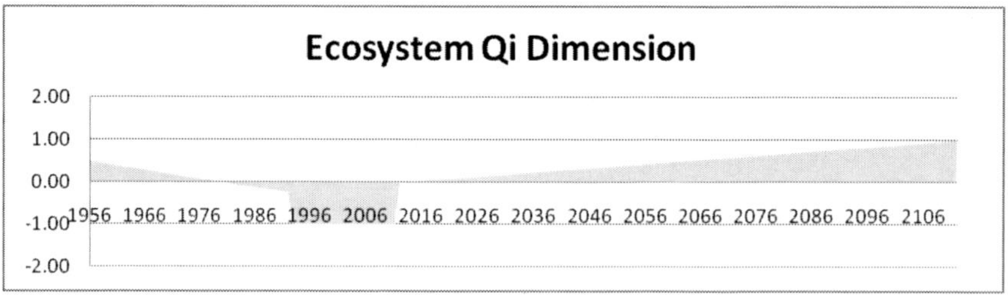

Figure 5. Cumulative Impact of Three Gorges Project Ecosystem Qi Dimension (Environmental).

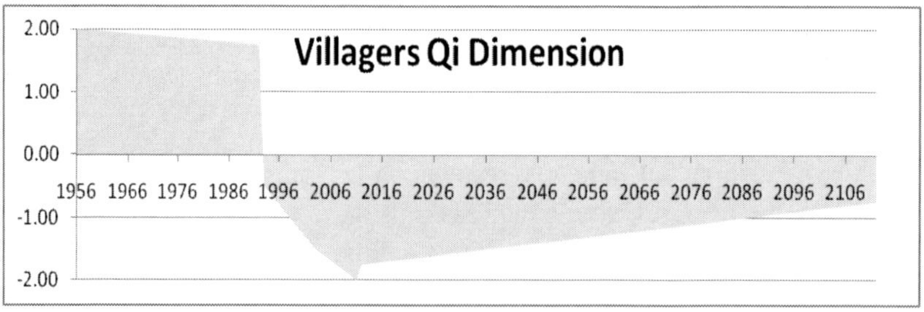

Figure 6. Cumulative Impact of Three Gorges Project on Villagers Qi Dimension (Cultural).

Table 4. Dimension Qi Averages for Sensitivity Analysis

<table>
<tr><th>Dimension</th><th colspan="2">1956</th><th colspan="2">1992</th><th colspan="2">1993</th><th colspan="2">2003</th><th colspan="2">2012</th><th colspan="2">2112</th></tr>
<tr><td rowspan="4">Ecosystem</td><td>+2</td><td rowspan="4">+0.5</td><td>+1</td><td rowspan="4">-0.25</td><td>-1</td><td rowspan="4">-1</td><td>-1</td><td rowspan="4">-1</td><td>+1</td><td rowspan="4">0</td><td>+1</td><td rowspan="4">+1</td></tr>
<tr><td>+1</td><td>-1</td><td>-1</td><td>-2</td><td>-2</td><td>+1</td></tr>
<tr><td>+1</td><td>+1</td><td>0</td><td>-2</td><td>-1</td><td>0</td></tr>
<tr><td>-2</td><td>-2</td><td>-2</td><td>+1</td><td>+2</td><td>+2</td></tr>
<tr><td rowspan="4">Village</td><td>+2</td><td rowspan="4">+2</td><td>+2</td><td rowspan="4">+1.75</td><td>+1</td><td rowspan="4">-0.5</td><td>-1</td><td rowspan="4">-1.5</td><td>-2</td><td rowspan="4">-1.75</td><td>-1</td><td rowspan="4">-0.75</td></tr>
<tr><td>+2</td><td>+1</td><td>-2</td><td>-2</td><td>-2</td><td>-1</td></tr>
<tr><td>+2</td><td>+2</td><td>0</td><td>-1</td><td>-2</td><td>0</td></tr>
<tr><td>+2</td><td>+2</td><td>-1</td><td>-2</td><td>-1</td><td>-1</td></tr>
<tr><td rowspan="4">Community</td><td>+1</td><td rowspan="4">0</td><td>+1</td><td rowspan="4">+0.25</td><td>0</td><td rowspan="4">-0.25</td><td>-1</td><td rowspan="4">-0.5</td><td>+1</td><td rowspan="4">+0.75</td><td>+2</td><td rowspan="4">+1.75</td></tr>
<tr><td>0</td><td>0</td><td>-1</td><td>-1</td><td>-1</td><td>+1</td></tr>
<tr><td>+1</td><td>+1</td><td>+1</td><td>-1</td><td>+1</td><td>+2</td></tr>
<tr><td>-2</td><td>-1</td><td>-1</td><td>+1</td><td>+2</td><td>+2</td></tr>
<tr><td rowspan="4">Family</td><td>-1</td><td rowspan="4">-0.5</td><td>-1</td><td rowspan="4">-0.75</td><td>-2</td><td rowspan="4">-0.75</td><td>-2</td><td rowspan="4">-0.25</td><td>+1</td><td rowspan="4">+1.75</td><td>+2</td><td rowspan="4">+1.75</td></tr>
<tr><td>0</td><td>-1</td><td>-1</td><td>-1</td><td>+2</td><td>+2</td></tr>
<tr><td>-1</td><td>-1</td><td>-1</td><td>+1</td><td>+2</td><td>+2</td></tr>
<tr><td>0</td><td>0</td><td>+1</td><td>+1</td><td>+2</td><td>+1</td></tr>
<tr><td>Overall Qi</td><td colspan="2">+0.50</td><td colspan="2">+0.25</td><td colspan="2">-0.63</td><td colspan="2">-0.81</td><td colspan="2">+0.19</td><td colspan="2">+0.94</td></tr>
</table>

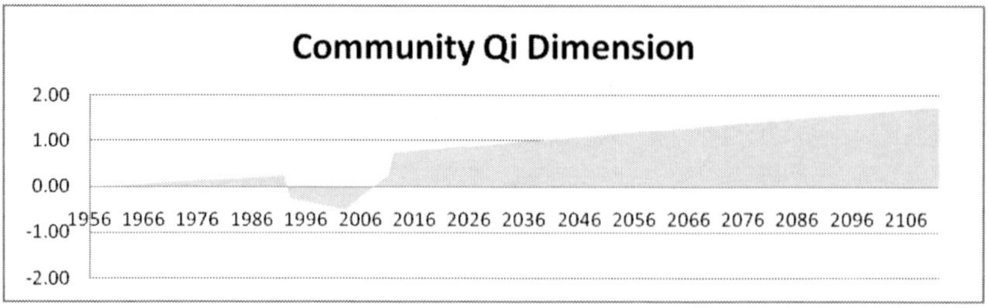

Figure 7. Cumulative Impact of Three Gorges Project on Community Qi Dimension (Social).

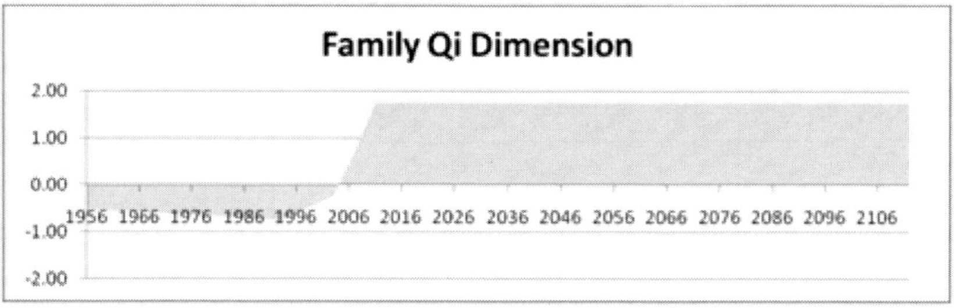

Figure 8. Cumulative Impact of Three Gorges Project on Family Qi Dimension (Economic).

The environmental and social impacts represented as changes in ecosystem and community qi, indicate increased stresses on both the fabric of the ecosystem and the fabric of society.

With the completion of the project however, these are both then enhanced over time. The cultural impacts reflect the absolute irreversibility of many of the changes that diminish qi associated with traditional occupation of the directly affected areas. The prior state of qi in this regard does not recover during the study timeframes and reflects the most substantially impacted dimension for the project. The unrealised economic opportunities of TGP are reflected in the changes of family qi, for the commitment of resources over a significant period to facilitate TGP, but producing significant economic returns for the region and China once the project is complete.

It is important to realise that different people hold different perspectives on the sustainability of the Three Gorges Project. Their perspectives are influenced most by the factors considered important in their own lives. Cultural groupings often hold the same worldview within their group, due to the way they identify with common values and beliefs. Indigenous people tend to consider cultural identity and ecosystem integrity the most important dimensions while bankers are more likely to focus almost entirely on the economic viability of a project. This bias can be acknowledged and accounted for in the MMDMF by incorporating the AHP results as a sensitivity analysis. The AHP results used for the sensitivity analysis are provided in Table 5.

Table 5. Analytic Hierarchy Process Results for Contemporary Worldviews

Dimension Priority	Chinese Government	Chinese Villagers	Chinese Other	Western (NZ) Worldview	Indigenous Worldview
Environmental Well-being	19%	17%	17%	19%	40%
Cultural Well-being	11%	17%	14%	9%	30%
Social Well-being	25%	33%	30%	33%	20%
Economic Well-being	45%	33%	39%	39%	10%

Table 6. Sensitivity Analysis for Three Gorges Project Event Years 1956, 2003, 2012 and 2112

Absolute Qi	+0.5	-0.81	+0.19	+0.94
TGP Event Year	1956	2003	2012	2112
Chinese Government	+0.09	-0.59	+0.79	+1.34
Chinese Villagers	+0.26	-0.67	+0.53	+1.20
Chinese Other	+0.17	-0.63	+0.67	+1.29
Western (NZ) Worldview	+0.08	-0.59	+0.78	+1.39
Indigenous Worldview	+0.75	-0.98	-0.20	+0.70

The Three Gorges Project AHP analysis produced similar priorities for the dimensions for all three groupings; Chinese Government, Chinese Locals and Chinese Others. The similar results could be accounted for in that the groupings represent the same cultural peoples, who continue to hold similar worldviews, but with slightly differentiated points of view. Pragmatically, these worldview groupings all consider the economic and social dimensions of Qi to be the most important, however while cultural and environmental are both weighted less, they are still acknowledged.

The surveyed 'Western perspective' differs, only in that the cultural dimension is given less priority. A significant influence and motivator in western thinking is monetary outcomes, reflected in the AHP results. Cultural wellbeing is not a dimension valued highly, perhaps due to the fact that different traditions and behaviours apparent in other societies are difficult to interpret and truly value from an economic standpoint.

The priority weightings are combined in Table 6 with the dimension averages taken from Table 4 for the TGP Event years 1956, 2003 following the completion of phase II construction, 2012, and for 2112.

DISCUSSION

The results obtained from the Qi-ometer analysis predict positive results for the sustainability of TGP. There is some polarisation in the sensitivity analysis, however within 100 years, all worldviews acknowledge that TGP will have made a positive contribution to China's sustainability. In 1992the decision was taken to implement the concept and construct TGP. The indicators chosen for the analysis reflect the broader research available today. Some of the research now available recommends indicators that were not considered in the 1988 CYJV feasibility study. For example the indicator for Environmental Geology is now considered due to the impacts of the water reservoir on the geology in the surrounding area was never predicted to be as severe as some geologists today believe it to be causing landslides.

Some general comments are made on the indicator assessment results. The Physical Impacts were seen to be positive in the analysis with the exception of the construction period. This is largely due to the massive amounts of carbon dioxide emissions that will be reduced and the ecosystem enhancements associated with or facilitated by the project. The Yangtze Dolphin has become extinct since the construction of the dam, and many other wildlife and plants have been impacted. The dolphin was considered endangered in 1988 and it was predicted that the species would eventually become extinct with or without the construction of TGP. Extinction of the Baiji has nevertheless been attributed to TGP as a negative impact on the environment. Slope stability and erosion have become an issue for the Hubei Province which geologists believe is an impact of the dam. The impacts of the huge concentration of water in the reservoir basin on localised climate change have been cause for concern and are linked to numerous landslides in the area.

TGP and the flooding of the reservoir basin have resulted in the loss or destruction of many historic relics. There has been wide criticism of the Chinese Government on the issue of relocating these artefacts to higher ground, before rising water levels flooded their original sites. The indicator is therefore qi diminishing for cultural and historical relics. The indicator Way of Life has been negatively impacted due to changes associated with both the relocation of around 1.5 million people and the altered context of those people's existence. Impacts on the Regional Identity are portrayed as qi diminishing. This is conservative as while the due to the issues surrounding sense of belonging are important, TGP has also provided a source of pride for the People's Republic of China. Satisfaction/Strain considers whether people are better off. Acknowledging that this is an extremely subjective measurement, and that the levels of happiness within the effected communities are difficult to comprehend as an 'outsider', the conservative approach is to assume that the impact is qi diminishing but note that some research suggests that TGP creates enhanced opportunities for the future and will potentially become qi enhancing. Economic indicators and associated industrialisation are the main motivation for TGP. These indicators are all positive and essentially TGP would not go ahead if that was not the case. Economic indicators are the primary focus in most development projects and are often most easily evaluated. TGP is now providing significant economic benefits for the region, and much of China.

The predictions for 2112 are all positive. The sensitivity analysis suggests all worldviews consider TGP to be enhancing qi, leading the modern world in adopting cleaner, greener power sources. China is a country with huge potential for more dams, due to the suitability of its topography. New Zealand is similarly disposed however the combined complexity of cultural, social and environmental issues associated with hydro-development projects is very difficult to overcome.

The MMDMF is an easy tool to implement in any project decision making. It is ideal for identifying areas of conflict. When working with people who come from different cultural backgrounds to one's self often it is easy to be naive and overlook issues that may appeal or conflict with others beliefs and values. The MMDMF identifies areas of concern, which can then be acknowledged, further investigated, and negotiated if necessary. As the MMDMF is intended to increase understanding in the decision making process, this can ease possible tension or contribute to resolving issues that could manifest later during the course of the project. The MMDMF is a useful decision making aide, providing greater focus on social and cultural impacts which are often immeasurable in other analyses. How does one establish the monetary value of lives saved from future floods? Many indicators often cannot be valued effectively in monetary terms, and although identified in feasibility studies, many are not included in the Cost/Benefit Analysis approach.

Finally the findings of this study are consistent with are search project (Morgan et al., 2012) that investigated the Cost Benefit Analysis based feasibility study that provided the justification for proceeding with the Three Gorges Project by comparison with an MMDMF assessment conducted on the same information available at the time that the decision was made. While the time frames for analysis differ, the results are consistent providing confidence in the findings of the previous investigation.

Conclusion

The combination of historic relevance, impacted anthropological sites, ecosystem modification, altered river flows, industrial and urban uses, power generation, transportation reliability, geological instability, siltation issues, and economic base, all under the Western World's spotlight, are some of the inter-related attributes that make the Yangtze River a complex management challenge: A contemporary interpretation perhaps of its metaphor; the Golden Waterway.

The Mauri Model Decision Making Framework offers a new approach to the analysis of water resource management as it provides a transparent and inclusive approach to considering the environmental, economic, social and cultural aspects of decisions that have been taken regarding the Yangtze River system. The Mauri Model Decision Making Framework is capable of understanding multiple-worldviews and adopts Qi in this case in the place of monetary equivalents as the base metric allowing an absolute sustainability assessment.

There are interesting parallels between the Yangtze River and the Waikato River in New Zealand, as while the actual physical scales are demonstrably different, the contextual relevance of the rivers resonates, as does the strategic importance of these rivers to their respective governments. The changing perceptions of both rivers from different worldviews, offers new ways of understanding sustainability, and better understanding the cultural and environmental implications of the development taking place on the Yangtze River. The application of the Mauri Model Decision Making Framework to the

Yangtze River reveals the parallels between the concepts of Mauri and Qi, and the potential opportunities for the application of the framework in other contexts of Asia and the Pacific.

REFERENCES

Bristow, M. 2007. Deep Concern over Three Gorges Dam. BBC News, Beijing [Online]. *http://news.bbc.co.uk/ 2/hi/asia-pacific/7120856.stm*. [23 May 2009].

Cao, S., Liu, X., Huang, E. 2010.Dujiangyan irrigation system – a world culture heritage corresponding to concepts of modern hydraulic science. JHER, 4. 3-13. Doi:10.1016/j.her.2009.09.003.

Chai, M.L. 2007. China A to Z: Everything you need to know to understand. London: Penguin Books Ltd.

Cipm. 1988. Three Gorges Water Control Project Feasibility Study: Volumes 1-11. Canada: CIPM.

Environmental impact assessment for developing countries. Tycooly International, London. p. 219–228.

Jackson, S. and Sleigh, A. 2000. Resettlement for China's Three Gorges Dam: socio-economic impact and institutional tensions. *Communist and Post-Communist Studies*, 33: 223-241.

Jin, L. 2009. Three Gorges on the Yangtze River in China. Hainan Publishing House.

Mccabe, A. 2007. Expert raises alarm about massive dam. Ottawa Citizen [Online]. *http://www2.canada.com/ottawacitizen/news/story.html?id=63ea413c-3a60-4177-9f92-58045ac1b996*. [23 May 2009]

Michael, T. 2005. The Pristine Dao – Metaphysics in Early Daoist Discourse. United States of America: State University of New York Press.

Morgan, T. K., Sardelic, D. S., and Waretini, A. F. (2012). Three Gorges Project: How Sustainable. *Journal of Hydrology*, 460-461, 1-12. doi:10.1016/j.jhydrol.2012.05.008.

Morgan, T.K.K.B. 2006.Decision-support tools and the indigenous paradigm. *Engineering Sustainability*, 159(4): 169-177.

Morgan, T.K.K.B. 2008. The Value of a Hapū Perspective to Municipal Water Management Practice; Mauri and its Potential Contribution to Sustainability Decision Making in Aotearoa New Zealand. *Civil Engineering*. New Zealand: University of Auckland.

Morgan, T.K.K.B. 2009. Takarangi, Yin and Yang, Mauri and qi. *MAI Review*, 2009(3): 1-3.

Needham, J. 1981. *Science in Traditional China: Comparative Perspective*. The Chinese Press, Hong Kong.

Qigang, D. 1998. What are the Three Gorges resettlers thinking? In Qing (Ed) The River Dragon has Come! IRN Barkeley.70-89.

Qing, D. 1998. The danger to historical relics and cultural antiquities in and around the Three Gorges area. In Qing (Ed) The River Dragon has Come! IRN Barkeley.124-142.

Qu, G.P. 1987. The role of environmental impact assessment in economic development.

Saaty, T.L. 1980. Analytical Hierarchy Process: Planning, Priority Setting, Resource Allocation. United States of America: McGraw-Hill.

Saracino, D.E. 1997. Environmental/Social Cost-Benefit Analysis: The Integration of Sustainability – The Cases of the Narmada Valley Project and the Three Gorges Project – A Thesis in The Department of Sociology and Anthropology. Canada: Concordia University.

Wang, Y., Morgan, R.K., Cashmore, M. 2003. Environmental impact assessment of projects in the People's Republic of China: new law, old problems. *EIA Review*, 23(5), p 543-579, ISSN 0195-9255, 10.1016/S0195-9255(03)00071-4.

World Commission on Environment and Development, (WCED). 1987. *Our Common Future*. Oxford University Press, New York.

In: Yangtze River ISBN: 978-1-62618-285-1
Editors: T. B. Maloney, B. R. Hutchins

Chapter 4

INNOVATIVE SOLUTIONS FOR YANGTZE RIVER'S WATER CRISIS

Shu-Qing Yang*

School of Civil, Mining and Environmental Eng.,
Univ. of Wollongong, NSW, Australia

ABSTRACT

Yangtze River is one of the largest rivers in the world, it has disasters like floods, droughts, sedimentation and water pollution etc. After the operation of Three-Gorge-Dam (TGD), the largest dam in the basin, the water shortage problem becomes very severe and frequent, especially in dry seasons. This chapter's analysis shows that the application of SPP scheme in Dongting and Poyang Lakes in the river basin can successfully eliminate these disasters. The so called SPP stands for "separation, prevention and protection", the chapter presents how to apply the SPP to these lakes. The feasibility is discussed in terms of technology, economical affordability and environmental sustainability. The chapter also compares the SPP scheme with the scheme of sluice gates/dam at the lake's outlets which the local governments have been put forward, the results show that the payback periods of two schemes are almost same, but the SPP scheme is much better than the gate/dam scheme in every aspect. It is found that the disasters in the river basin can be effectively eliminated if the proposed SPP is applied in the basin's lakes.

* E-mail: shuqing@uow.edu.au.

Keywords: Dongting Lake, Poyang Lake, Yangtze River, SPP scheme, dam scheme

INTRODUCTION

Yangtze River: is the longest river in Asia, and the third longest in the world. Its flow path is as long as 6,418 km from the glaciers on the Qinghai-Tibet Plateau eastward towards central and eastern China before draining into the Eastern Sea at Shanghai at the annual runoff to nearly 1,000 billion m^3 (see Table 1) that is the fourth biggest river by discharge in the world. The river drains one-fifth of the land area of China (1.8million km^2) and its river basin is a home to 400million population, especially in several large cities like Shanghai, Nanjing, Wuhan, Chongqing, Chengdu, Changsha and Nanchang etc.. This is a rich basin because with only 20 % of the nation's total land area, more than one-third of China GDP is produced, its gross industrial and agricultural output occupies about 40% of the whole nation. The Yangtze Basin, particularly in the middle and lower reaches, is one of the most socially and economically developed regions in China. The prosperous Yangtze River Delta generates as much as 20% of the China's GDP.

The upper course of the river originates in glacial meltwaters on the slopes of the Tanggula Mountains and descends through deep valleys in the mountains east of the plateau till Yichang where the Three-Gorges Dam locates (see Figure 1). The River forms a narrow valley up to 3 km in depth. Individual mountain peaks exceed elevations of 4,900 m above sea level and are crowned with glaciers and perpetual snow. The steep, rocky slopes are cut with gorges and deep valleys.

The middle course of the Yangtze stretches for 955 km between TGD and the outlet of Poyang lake. The climate is characterized by hot summers and relatively mild winters. Annual precipitation measures between 1,000 to 1,500 mm, a large part of it occurring in summer; the wet season lasts for more than six months.

The river widens markedly the course of its stream wandering in the form of a large loop. The banks along the middle and lower course are built up for protection from floods. In the southern part of the plain lies Dongting Lake and Poyang Lake. The former was the largest freshwater lake in China but now has been reduced in area by silting and land reclamation, and it shares four tributaries and three canals with the Yangtze, the plain of Dongting Lake is China's most important rice-producing region.

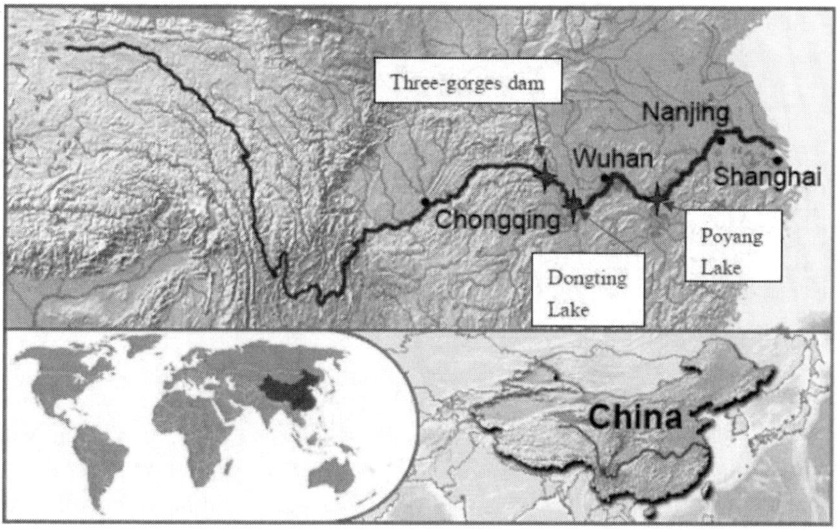

Figure 1. The Yangtze River and its important lakes and cities in the basin.

Lake Poyang is currently China's largest natural freshwater lake. The lake, with an average area of about 3,585 square km, receives water from the 5 river tributaries, in turn, is linked to the Yangtze by a wide tributary. The river then turns to the northeast, passes through a widening valley. From the Lake Poyang till its estuary, the river flows to its lower reach with 938 km long, which has suffered from industrial pollution, agricultural run-off, siltation, and loss of wetland and lakes, which exacerbates seasonal flooding (Yang and Liu, 2011). The banks have been strengthened after the 1998 floods; the total length of artificial banks with levees is 2,740 km. The river channel can safely discharge floods at 60,000-70,000 m^3/s - equivalent to 10 to 30 year frequency flood flow.

The average yearly rainfall in the basin is about 1,100 mm. 60-80% of the precipitation is brought by the monsoon winds and falls primarily as rain in the period May-Oct., which results in floods in the middle and lower parts of the basin. In 1870, 1931, 1954 and 1998, the severe flood disasters incurred the total life losses up to 3-5 million. If the 1870 flood happened again, the discharge could be as high as 105,000 m^3/s. The historical flood events were very useful for the people to support the proposal of Three-Gorges Dam.

Table 1. Water contributions to the main stream of Yangtze River by sources, unit billion m^3

Upstream inflow (Three-Gorge Dam)	448
Lateral inflows	127.5
Lake Dongting	166
Lake Poyang	146
Han River	68
Huaihe River	20.5
Total	976

Three Gorges Dam (TGD): is world's most powerful dam which was initially proposed Sun Yat-sen in 1919. The founder of P. R. of China, Mao Zedong initially supported the idea, but finally gave up due to the consideration of possible civil wars. The resources for the production of energy from the Yangtze are enormous, the total potential power is estimated to be more than 200 million kilowatts, representing about two-fifths of the total hydro-energy potential of all the rivers of China. The construction of the dam started on December 14, 1994, and started in operation in 2006. The dam controls a drainage area of 1 million km^2, the flood storage capacity is 22.15 billion m^3, and installed capacity 18,200 MW, yearly output 84.7 billion kWh. The dam can also improve the navigation condition in the dry season in the middle reaches. Twelve new dams planned for the river will join Three Gorges Dam to address the country's energy needs, it also poses tremendous threats to river's unique ecosystems. The main problems are listed as follows.

Water crisis in Yangtze River Basin: the Yangtze River is still an unhealthy river, and the water crisis includes floods (too much), droughts (too little), sedimentation (too turbid), water quality deterioration (too dirty), wetland shrinkage and ecosystem crisis.

Before the construction of the Three-Gorges Dam, it was predicted that although the dam is very effective to control the floods, the excessive water in the area of Dongting Lake would still be 16-30 billion m^3 if the 1954 flood appeared again. It was also predicted that the droughts would be improved as the dam would release the stored water in dry seasons, but in reality after the TGD started its operation the water level in the river started to be lower and lower in dry seasons and droughts become worse. It was predicted that the large ships of up to 10,000 tons displacement can sail to the upstream like Chongqing, but now the large vessel is very hard to reach Dongting Lake due to insufficient water in the navigation channel in dry seasons. Table 2 lists the noticed crisis for the river and especially the lakes of Poyang and Dongting. Among them, the very critical problem is the water shortage. If no good measures are taken, it is predictable that the river will be cut off its flows in the main stream in the near future, the calculated results shown in Table 3.

In the natural conditions, the long-time averaged flow rate at TGD during October is only 19000m^3/s. After a flood season the Three-Gorges Dam needs to raise the water level and this reduces 8000m^3/s in the main stream from Sept. In order to eliminate the water shortage crisis in northern China like Beijing, Tianjin etc., China is currently constructing the world's largest water diversion projects from the Yangtze river to the north, the project will be completed in the near future, it is estimated that the project needs at least 2000m^3/s of water from the river in dry seasons. On the other hand, many large dams like Xiangjiaba, Xiloudu etc. are under development, and these reservoirs also need to store water from September, and the reduction of flowrate will be at least 6000m^3/s. At this moment, both Dongting Lake and Poyang Lake are proposed to construct dams at their outlets, at least 2000m^3/s of water from each lake should be reduced to the river if the proposal is adopted by the central government. Therefore the river flowrate at the downstream or estuary could be less than 1000m^3/s. It will have very negative impacts on the downstream people's life and ecosystem.

Table 2. Water crisis in Yangtze River

	Yangtze main stream	Dongting Lake	Poyang Lake
If 1954 floods occurred again	TGD can reduce the loss	Needs 16-30 billion m^3 for detention	Needs 11.7-12.7 billion m^3 for retention
If 2011 droughts occurred again	TGD reduces 8000m^3/s in dry seasons	No enough water, the lowest water level = 6.1m above sea level	No enough water, the lowest water level = 5.9 m above sea level
Water pollution	Poor	Deterioration	Reasonably ok
Sedimentation without TGD	17millon tons/year	131 million tons/year	10 million tons/year
Ecosystems	No. of Chinese river dolphin is being reduced	rat disasters outbreaks in 2003, 2005, 2007.	Birds reduction
Navigation	No enough water for navigation in dray seasons		

Table 3. Water in Yangtze River after the construction of hydraulic projects in dry seasons

Water available at TGD site before its operation	19000 m^3/s
TGD retention to increase its water level for hydro-power	8000 m^3/s
South-North Water Diversion project to be completed in 2014	2000 m^3/s
29 other upstream large dams (under construction/plan)	6000 m^3/s
Dongting Lake dam (under plan)	2000 m^3/s
Poyang Lake gate (under plan)	2000 m^3/s
Flows to the estuary	1000 m^3/s

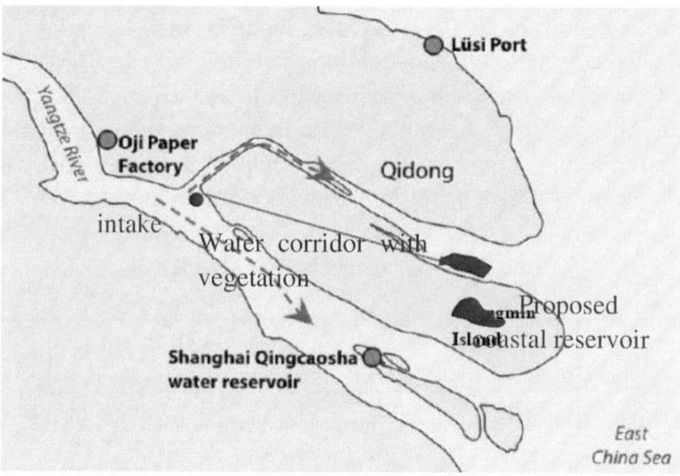

Figure 2. Coastal reservoir at Yangtze estuary-Qingcaosha and the reservoir proposed by the author in 2006 is highlighted in red.

The runoff reduction in the mainstream may have disastrous effect to Shanghai, the largest industrial and business city in China. In 2005, the author proposed a coastal reservoir (a freshwater reservoir in sea water) to Shanghai government as shown in Figure 2 when the city was faced with the water deficit due to pollution (Yang 2006).

In 2010, Shanghai completed the construction of coastal reservoir-Qingcaosha which is located at a place different from the author's suggestion. Yang (2006) suggested that the intake of coastal reservoir should be located at the upstream of Chongmin Island, which is represented by a dot in Figure 2, the reservoir should be located at the north Chongmin, which is denoted by the red area, the water should be transferred from the intake to the reservoir by a natural channel where the wetland could purify the river water, this is specially needed as the riverwater could be heavily polluted by upstream industrial, agricultural and domestic wastewater.

Unfortunately, the author's proposal was not adapted by the local government. Instead, they constructed the coastal reservoir-Qingcaosha in the south Chongmin, it opens the gates to take the river water during low tidal when the freshwater surrounds the reservoir, and its gates are closed during high tide when the seawater surrounds the reservoir. The author once clearly told the decision maker the shortcomings of the Qingcaosha reservoir as follows:

1) It is located inside the main flood passage. If 1870, 1954 floods happened again, this reservoir could be washed away in to the sea, or it has the high risk caused by too much water;
2) In dry seasons, the reservoir may have no freshwater to take when the seawater surrounds the reservoir by months even in low tide. This duration of intake failure could become longer and longer after the construction of South-North water diversion project and 29 other dams finish their construction, or it has the other high risk caused by too little water.
3) Algal blooms may appear in the reservoir, this is because that the non-point wastewater produced by cities like Shanghai may be collected by the reservoir as the tidal flow could push the wastewater into upstream, and eventually the nitrification could be noticeable, or it may have the problem caused by too dirty water.

If we compare the Qingcaosha and the author's proposal, one can easily conclude that all risks mentioned above could be avoided if the author's proposal is adopted. For example, the biggest floods would not wash the reservoir marked in red in Figure 2 into the sea; the intake locates at the upmost of Shanghai territory, thus it has the lowest possibility for failure intake of freshwater; the wastewater produced by Shanghai is unable to enter the proposed reservoir, and also the water from the intake can be further purified by the wetland on Chongmin Island as shown by the blue arrow line in Figure 2.

It can be seen that many hydraulic projects are undergoing in the Yangtze River basin, the Yangtze River may be a dead river without water in dry seasons if these projects and their management are not well planned. From the literature, no other paper already alerts this immediate crisis as shown in Table 3, i.e., Yangtze River will be a seasonal river. Thus correct actions are needed to take from now to save the Yangtze River.

Recently Yang and Liu (2011) proposed the SPP strategy to harness rivers and lakes, its core is that clean water and unwanted water (dirty turbid waters or safe floods) should be separated spatially and temporally; clean water should be stored and protected against pollution and wastewater should be discharged as fast as possible. In other words, the detention time of unwanted water in a river system should be as short as possible, whilst the retention time of wanted water should be as long as possible.

The goals of SPP strategy to eliminate the water crisis (too much, too little, too dirty and too turbid) could be achieved by constructing a by-pass channel (BPC) in a lake, which would divert low-quality water from the lake during low precipitation periods and allow better quality water to flow into the lake during high flow periods, and the clean water will be stored. This chapter we will investigate whether the mentioned SPP can solve all problems that Yangtze River has now.

DONGTING LAKE'S WATER CRISIS AND THE COUNTERMEASURES

Dongting Lake is an efficient flood regulator of Yangtze River and was the largest lake among thousands of lakes in the basin, but unfortunately it becomes the secondary largest as a result of the sedimentation and reclamation over the past 50 years. Its area was about 263,000km^2. Dongting Lake 's capacity in 1949 was 29.3 billion m^3 , but 44 years later it decreased to 17.4 billion m^3 in 1983, and adding to the reclamation it's area shrank from 4350 km^2 in 1949 to 2691 km^2 in 1983. Four rivers of Xiangjiang, Zishui, Yuanjiang and Lishui drain from south and west, besides three channels connect the lake to the Yangtze River, which dumps a major proportion of the sediment suspended from Yangtze River to the lake when the lake accommodates floodwater. It has only one outlet on the east side to discharge the water to Yangtze River (Fig. 3).

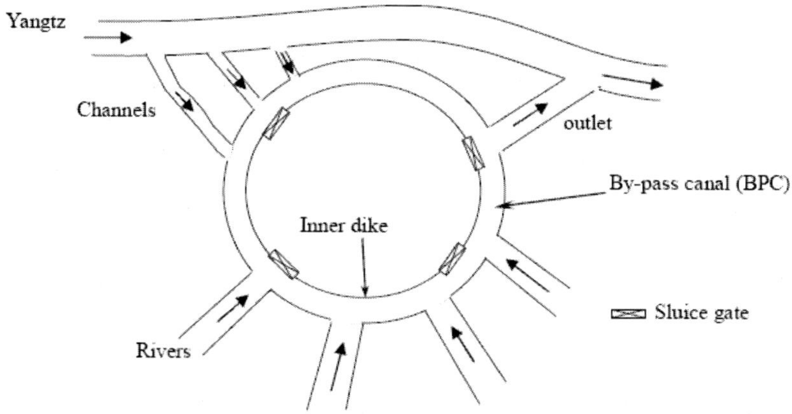

Figure 3. SPP Scheme and its By-Pass Canal (BPC) in Dongting Lake.

After the operation of TGD, drought disasters arise more and more frequently particularly in dry season, this becomes very serious. Subsequently someone conceives a measure by constructing a sluice gate at the outlet of the lake to store the water during dry seasons. Ostensibly it seems to regulate water temporally and ameliorate the water shortage in dry seasons, nevertheless it will exacerbate another problems including flood and sedimentation. If a sluice gate is installed to retain water, unfortunately the sedimentation and contaminant would be worser, and gradually the sedimentation would occupy its available capacity.

In order to mitigate the sedimentation, a lot of strategies have been proposed and implemented in Dongting Lake. The most traditional way is to heighten and strengthen the levees to increase flood storage capacity, in long term, this is only an expedient measure by hiding the danger.Recognizing this shortcoming, some experts claim that the core issue is to dredge lake's outlet channel. Due to sedimentation, the wet land area is shrinking gradually resulting in the ability lessening of the lakes both in water storage and flood regulation.

SPP Strategy: Generally lakes collect water from upstream rivers/creeks and drain to the downstream as shown in Figure 3. In Dongting Lake, there are three channels link the Yangtze River with the lake and four rivers from Hunan Province, the outlet rejoins to the Yangtze River at Chenglingji.

The structures needed by SPP strategy is to construct an internal levee with sluice gates around the lake shoreline as a by-pass canal (BPC) in Fig. 3, which can be assimilated to an artificial river with regulatory function. We will discuss how the water crisis could be eliminated by the SPP scheme.

1) *Sedimentation and pollution.* The analysis above demonstrates that the unwanted water comes from the upstream, thus the sediment and contaminant accumulated in lakes are the biggest killer for the storage capacity and clean water. Here with BPC, the incoming water is divided into wanted and unwanted water according to its applicability for human activities, the lake can separate the two kind of water with the aid of sluice gates on BPC in the following way:
   If the river water is heavily polluted or sediment discharge is very high, the sluice gates will be closed so that all polluted water by-passes the lake and is discharged to the downstream via the lake outlets. In flood seasons the water quality from upstream is relatively better as a result of the silt and nutrient concentration is very low except the first peak, so the sluice gates will be opened and the lake starts to store the clean water. In dry seasons, the sluice gates will be closed to prevent the unwanted entering the lakes. Compared with the natural lakes without BPC, the inner levee can resist the silt and pollutant so that the internal clean water can be protected. In another words, the BPC can manage the water in time and space according to water quality and human's need. This strategy can success mostly owe to the BPC where the cross-section area is much smaller relative to the lake, thus for a certain discharge at the outlet, the velocity in the PBC will be rather fast. Since the running water never becomes putrid, the relatively higher velocity ensures that it can keep its own good quality and protect the inner lake from being silt and polluted.
2) *Flood control.* For a natural lake, the first flush of the inflowing rivers during flood seasons always occupies the lake's available capacity, consequently the lake has no enough capacity to regulate the following flood peak, thereby the flood water brings about destructive disasters. However the SPP strategy can control the water's in and out intentionally, it discharges the first flushes in flood season by the BPC and increases the effective flood-control storage, i.e., the safe flow will not be allowed to enter the lake to occupy its capacity, the sluice gates are always closed not only to refuse the coming of wastewater but also to guarantee the sufficient capacity until the arrival of flood peaks that excess the designed water level. Consequently this SPP scheme can completely mitigate the flood disasters. Moreover, in a long run, this method will reduce the lakes' dead storage relatively

compared with natural situation as the BPC can discharge the majority of unwanted water to the downstream so that it enhances the lakes' ability of regulating flood and prolongs their life.

3) *Water supply*. Just like a common reservoir, when all the sluice gates are closed in dry season, water in the lake will not drain to the downstream by gravity and can stay high water level even in dry season. This strategy with BPC has the advantage as it stores clean water in flood seasons at first and prevents it from mixing with wastewater in dry season, therefore it can provide available clean water for human especially when the precipitation is quite little.
4) *Wetland protection*. In fact, human's reclamation is one of the serious reasons causing the lakes' decline. Residents make use of the shoal around the lakes constantly only for their immediate interest yet result in area shrinking and water pollution, though the government exhorts them all the time. Once the inner levee is constructed, reclamation will become quite difficult for the residents. In this degree, the SPP strategy can protect wetland from diminishing.

From the analysis above, the idea of SPP separates spatially and temporally the wanted water and unwanted water by the BPC based on the quality of incoming water and retains the clean water for a long time.

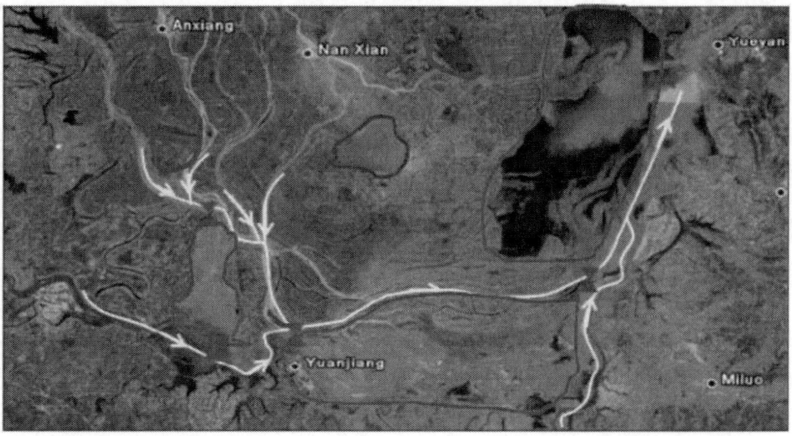

Figure 4. Application of SPP to the eastern, southern and western lakes where the red lines are the levees, yellow lines are the streams and the triangles are the sluice gates.

Consequently it can solve the five problems (too much, too little, too turbid, too dirty and wet land decrease) at the same time and doesn't bring about another new problem like other proposals. This could be achieved when the whole Dongting lake is treated as a lake system as shown in Figure 3. Alternatively, this can be achieved by treating it as the eastern, southern and eastern lakes system as shown in Figure 4, both need to investigate their feasibility.

POYANG LAKE'S WATER CRISIS AND ITS COUNTERMEASURES

Lake's water crisis: China has numerous lakes, among which the number of natural lakes with area above 1km^2 is more than 2800, the total water storage of all lakes amounts to 707 billion m^3, with the fresh water 225 billion m^3, accounting for 8% of the fresh water resources on the land. The lakes are not only valuable natural resources but also have many kinds of functions like flood disaster mitigation, navigation, irrigation, tourism attraction and fishery etc.

However, with the frequent human activities and climate changes, the number of the lakes has kept diminishing, with a total of 243 lakes covering more than one km^2 have disappeared in China over the past 50 years. Although China has about 24,000 natural lakes, they are disappearing at a rate of about 20 every year. Most lakes in China have also begun to suffer from various environmental stresses, including deterioration of its water quality with increasing nutrient and other chemical inputs. Lakes in the Yangtze basin have suffered from dropping water levels when so many dams and water diversion projects are undergoing, they also have been affected by problems such as water eutrophication due to rapid growth in economic and population, urbanization.

Poyang is a shallow and long lake where its average width is about 16.9km and its length from north to south is 173km, and it has a catchment area of 2923.7 km^2. Poyang Lake is now facing with two water quantity issues, i.e., too much water during high precipitation months, which causes flooding, and too little inflow during dry periods, which results in seasonal water shortages as lake level declines.

During the rainy season, floods often occur and cause heavy damage with overtopped water levels. While water level declines during the long, dry seasons leading to limitations in water supply. In May 2011 drought in Poyang Lake, the area of the water surface shrinks from 4000km^2 to 50km^2, and the water level was as low as less than 8m above the sea level. From April to September, the water resource in the lake is so rich that its highest water level could be 22.59m over the years and the water surface area in the lake could reach to 4647km^2. While during the rest time, the area of the water shrinks a lot and the lake almost can be seen as a river. With this large difference between supply and demand, a major proportion of the water used by industry, agriculture, and urban areas is discharged into the water system; there generally is a deterioration in water quality. Thus, there is a need to develop a hydrological solution to control floods while maintaining a sufficient quantity of high quality water year round.

Figure 5 depicts the pollution sources of freshwater in a lake. Among all those pollution sources, the non-point pollution is very hard to control. As lakes have little flow and contain stratified layers, they are vulnerable to pollution. It is very difficult to control pollutants from nonpoint sources. Water quality is deteriorating, sediment is accumulating, wetlands are shrinking in size and aquatic organisms are dying out.

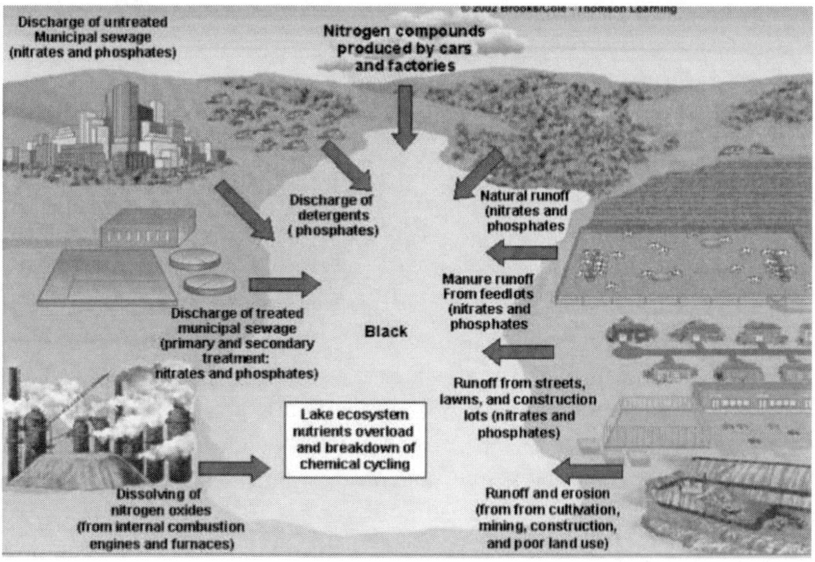

Figure 5. Pollution of freshwater lakes (http://courses.cqjtu.edu.cn/jpkc/hjkcxfz/en/show.aspx?id=147andcid=35).

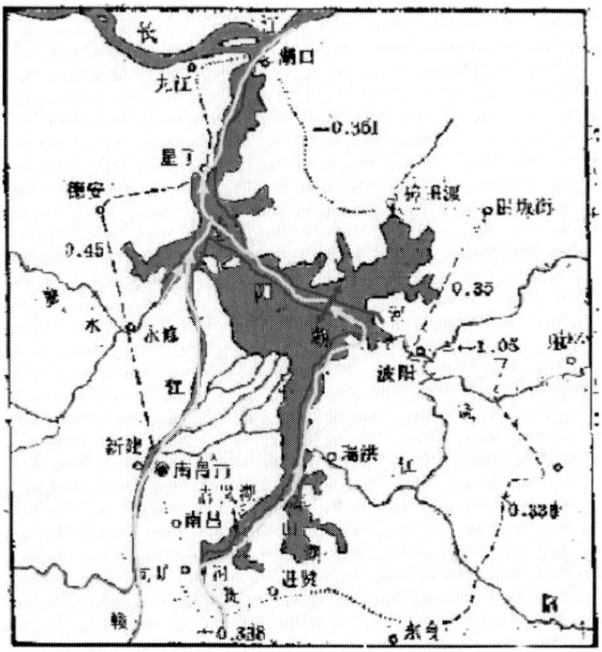

Figure 6. River system of Poyang lake and the application of SPP in which the sympols of inner levee/sluice gates are the same as that in Figure 4.

Consequently, the lake's capacity for flood control and drought relief has been affected. In natural conditions without separation, prevention and protection, the sediment is settled in the lake, and contaminant is mixed with the clean water, the floodwater under safe level occupies the capacity, thus all problems appear.

SPP application in Poyang Lake: Wanted and unwanted water should be separated spatially and temporally, the detention time of unwanted water in the lake should be as short as possible, whilst the retention time of wanted water should be as long as possible. To achieve the above goal, a by-pass channel (BPC) could be constructed, which would divert the unwanted water from the lake during low precipitation periods and allow wanted water (high quality) to flow into the lake during high flow periods.

The BPC can bypass the unwanted waters with high pollutant, high turbidity, only clean floodwater will flow to lake, thus there is no pollutant is introduced into the lakes and the water body is perfectly protected. The SPP required levees and gates are shown in Figure 6.

In Figure 6, the inner dike to be constructed is represented by red lines, and the sluice gates are denoted as red double-arrows. The BPC can isolate the wanted water from the unwanted water, the incoming flow can be separated into wanted/unwanted water based on its water quality and quantity. The most effective way to treat polluted water in a lake is to maintain water movement since moving water has the capacity of self-degradation, self-decomposition and self-purification, and high turbulence shears disperse planktonic flocs and limits sessile microbial growth. Now, more attention should be paid to the water shortage in drier months, because the water level is so low that the ecosystem suffers the disaster.

Now if we constructed the BPC in Poyang Lake, as it explained before the sluice gates would be opened during April to June to store the large amount of clean water, and then in dry seasons they would be closed to prevent the polluted water by discharging them to the downstream. The apparent advantage of this measure is that the BPC breaks the non-uniformity of water storage, i.e., it stores large quantity of water when the river water quality is quite well, it discharges the small quantity of river water when its quality is poor. If so, there will be enough water storage to cope with the water shortage, and then mitigate the drought threat. Some experts intend to construct an sluice/dam gate in the outlet of the lake, thus it is useful to investigate the difference between SPP scheme and the scheme of sluice/dam at the outlet .

ANALYSIS OF SPP EFFECTS TO YANGTZE RIVER SYSTEM

As mentioned, in natural conditions, Dongting Lake and Poyang Lake receive a lot of water and sediment. For the Dongting Lake alone, it receives about 300 km^3/year of water from the Three Channels and Four Rivers, about 1/3 of runoff in the Yangtze River. Among them, 41.4% of it comes from the Yangtze River via the Three Channels, and the rest mainly from the Four Rivers. Data shows that the Yangtze is responsible for 70-75% of China's floods, most of them occur in areas surrounding the Dongting Lake. In 1998, the flood resulted in a total loss of 4150 people, and 180 million people were affected, a staggering 100,000 km^2 were evacuated and 13.3 million houses were damaged or destroyed. Dongting Lake annually receives 1730 million tons of sediment, 83% of which is carried into the lake from the Yangtze River, and 17% from the Four Rivers. Over the past 150 years, the lake area has reduced from about 6000km^2 to about 2600km^2 due to both natural siltation and human activity, consequently flood modulation of the lake capacity has been reduced significantly from about 30km^3 in 1949 to 17km^3 in 1995. There is no doubt that if no measures are taken to mitigate the sedimentation rate, Dongting Lake would disappear in the near future.

The water crisis of floods, water pollution, sedimentation has existed for long time, but the operation of Three-Gorge Dam and other new dams together with the South-North water diversion project worsens the water crisis- the droughts will be very severe and frequent. To solve the water shortage problem, dams at the lakes' outlet have been proposed, and their feasibility study is being undertaken. It is useful to compare the different schemes between the sluice/dam proposal and the SPP scheme, and the results are shown in Table 4.

In Figure 4, the sluice/dam scheme at Poyang Lake will start to store water in Sep. and the Dongting Lake will do so in Oct. every year, this is the same period as the Three-Gorge Dam that significantly reduces the Yangtze River flow from Sept., thus the dam scheme worsens the droughts in Sept-Dec. According to the master plan, the dam water level at Poyang Lake will be 15.5m above the sea level, and the capacity will be to 10-20 billion m^3; the dam level at Dongting Lake will be 25m and storage capacity will be 3 billion m^3. However, the SPP will store the floodwater in June-August, before the Three-Gorge Dam's reduction in flow.

If design properly, each lake should have storage capacity up to 30 billion m^3. Obviously, SPP has the capacity to increase water to the downstream in Sept.-Dec. in order to quench the thirsts, to enhance the water level in the Yangtze River for navigation and to ensure the intake of Qingcaosha coastal reservoir for Shanghai's water supply.

The sluice/dam scheme has nothing to do flor disaster mitigation if 1954 and 1998 floods appeared again, but SPP will completely eliminate the flood disasters. Therefore, it can be seen that SPP scheme is a better solution for disasters mitigation caused by floods and droughts relative to the sluice/dam scheme.

The other problem in the Yangtze River is caused by sedimentation. In natural condition, about 50% of sediment laden by 146 billion m^3 runoff deposits in the Poyang Lake, and 70-80% of it by 300billion m^3 of water in Dongting Lake. If the sluice/dam scheme was adopted in these lakes, it is natural that the sedimentation rate must be higher than that in natural conditions. However, the SPP scheme can significantly reduce the sedimentation rate. If the sediment concentration is C, the sedimentation volume by its storage water (30billion m^3) would be 30C, and the total incoming sediment volume by 146 billion m^3 water would be be 146C, and the rate of sedimentation for Poyang Lake can be calculated as:

$$\frac{30C}{146C} = 20\%$$

It means that about 80% of incoming sediment will be discharged to the downstream via BPC, and only 20% of incoming sediment deposits in the lake. Similarly the sedimentation rate of Dongting Lake after the application of SPP will be

$$\frac{30C}{300C} = 10\%$$

It means that about 90% of sediment will be discharged to the downstream via the BPC, and only 10% of incoming sediment deposits in the Dongting Lake. Obviously, the sluice/dam scheme magnifies the shrinkage rate of wetland area, but SPP is useful to protect the wetland.

Table 4. Comparison of sluice/dam scheme with SPP scheme in terms of sustainability and affordability

	Sluice/Dam scheme at outlet		SPP scheme	
	Poyang	Dongting	Poyang	Dongting
Droughts in Sept.-Dec.	Water level 15.5m, 10-20b m^3 capacity to store water in Sept.	Water level 25m, 3b m^3 capacity in store water in Oct.	30b m^3 capacity, to store peak floodwater.	30b m^3 capacity to store peak flood water
Floods in June-Aug.	No capability to mitigate disasters	No capability to mitigate disaster	To mitigate 1954 floods	To mitigate 1954 floods
Sedimentation	70-80% of incoming sediment/year	70-80% of incoming sediment/year	20% of incoming sediment /year	10% of incoming sediment/year
Wetland	High shrinkage rate due to higher sedimentation rate		sedimentation rate is reduced and the shore is stablized	
Construction cost (billion yuan)	8	12	120	120
Payback period (years)	1.07	8	8	8

It would be interesting to compare the construction cost and its payback period. It was estimated that the costs for sluice/dam schemes at Poyang Lake and Dongting Lake are 8 billion and 12 billion Yuan, respectively. The SPP scheme has relatively higher construction cost and it is estimated that 120 billion Yuan could be enough to construct the levees and sluice gates. If we assume that 1 m^3 water can produce 0.5 Yuan benefit, then every year the benefit would be 15 billion m^3 of water *0.5 yuan/m^3, thus the payback period of 8 billion yuan investment of sluice/dam scheme for Poyang Lake (15 billion m^3 water/year) will be

$$\frac{8billion}{15*05} = 1.07year$$

Similarly, the payback period of sluice/dam scheme for Dongting Lake will be

$$\frac{12billion}{3*05} = 8year$$

But the payback period of SPP scheme for both Poyang and Dongting will be

$$\frac{120billion}{30*0.5} = 8year$$

It can be seen that there is no difference in terms of payback period between the SPP scheme and the sluice/dam scheme in Dongting Lake.

CONCLUSION

With the rapid growth of population together with agricultural and industrial development, people's demand for water quantity and quality would be increasing gradually. Asa flood regulator and water provider, lake plays an important role to protect the downstream region and ensure enough water supply for people's activities. However, there exist five problems including sedimentation (too turbid), flood disaster (too much), drought (too little), pollution (too dirty) and wetland decrease. If there is no effective strategy to manage our water resources, the modern civilization may collapse that had happened in the history. This chapter proposes the SPP water management strategy and discusses its application in lakes along Yangzi River. We can conclude that

1) The analysis shows that the water must be managed for its quality and quantity simultaneously. In other words, the appropriate water management must include particle management or impurity management;
2) Currently the water has been separated into many groups, like wastewater, river water, rainwater, floodwater etc. It is suggested that the water can be further separated as wanted and unwanted water based on the purpose of water management. Generally, the unwanted water includes heavily polluted or highly turbid water, or safe water, and the wanted water could be excessive floodwater or clean river water;
3) The SPP strategy refers to water separation, clean water protection and prevention from external pollution. Currently in the world, all reservoirs and lakes' water are exposed to external pollutants without protection, and the incoming river water has never been separated. Consequently, all clean lake/reservoir water has been polluted and the storage capacity is quickly lost by sedimentation.
4) After the application of SPP in Dongting lake and Poyang Lake, it is effective in solving the five problems , and all disasters can be mitigated with the help of SPP strategy.
5) In this book chapter, we only outlined the application of the SPP to the mentioned lakes, but in principle, the strategy is valid to all other large water bodies in the world. Next, more detailed research like physical and/or numerical models are needed before it can be used in practice. Eventually, it is subject to the decision maker/politician's decision.

ACKNOWLEDGMENT

The works are supported, in part, by National Science Foundation of China (51228901 and 51239001) and special fund by the Ministry of Water Resources. Some data used in this chapter was collected by a student from Sichuan Univ., China.

REFERENCES

Yang S. Q. and Liu P.W. (2010). "Strategy of Pollution Prevention in Taihu Lake and its Effects analysis". *J. Great Lakes Research*, 36(1), 150-158.

Yang, S.Q. (2006). "*Completely Solve Water shortage Problem in Shanghai by Coastal Reservoirs*". *http://www.paper.edu.cn*.

Yang, S.Q. (2004). "Global and China's water crisis and its solutions in the 21st Century". Tianjin Univ. Press, China.

In: Yangtze River ISBN: 978-1-62618-285-1
Editors: T. B. Maloney, B. R. Hutchins

Chapter 5

THE RESEARCH OF ENVIRONMENTAL FLOWS METHODOLOGY IN RIVERS OF CHINA

X. Ban
Key Laboratory of Monitoring and Estimate for Environment and Disaster of HuBei Province, Institute of Geodesy and Geophysics
Chinese Academy of Sciences, Wuhan, China

ABSTRACT

Environmental Flows (EFs) (Troldborg et al.) refers to the basic flow rate and process retained in the river for maintaining the river ecosystem. At present, there are many methods to determine the EFs, so it has become a frontier and hotspot issue now to choose the right EFs suitable for the characteristics of rivers in China. There are many rivers in China, with great difference in natural environment and ecosystem. Especially due to different levels of human development and utilization, it is difficult to determine the EFs with uniform standards, and different measures should be taken based on the difference in rivers and ecological protection goals. Based on the idea of river zoning and classification, by using research achievements at home and abroad, and based on different ecosystems of rivers in China, this paper classifies the EFs problems into three spatial levels: whole country, river basin and water function. Based on the characteristics of rivers in different regions and different reaches and in combination with ecological and environmental problems, different methods are used to calculate and determine EFs. This paper comprehensively summarizes the research of EFs in China, and recommends using different EFs calculation methods based on the characteristics of rivers in China, aiming to provide a scientific guidance for the research in China.

INTRODUCTION

EFs refer to the basic flow rate and process retained in the river for maintaining the river ecosystem, including: estimation of minimum flow rate and the flooding process in flood season; water flow process, hydrodynamic process as well as physical and chemical change process (Smakhtin 2007). At present, there are many methods to determine the EFs, more than 200 known. To sum up, they can be classified into hydrology method, hydraulics method, habitat simulation method and holistic method (Tharme 2003). The results calculated by various methods are different, and sometimes with larger difference. It is disputed how to determine the EFs of a river. As for the methods suitable for rivers in China, it often depends more on the river's current situation and the ecosystem in the future. The domestic understanding of evolution and value of river ecosystem is relatively shallow. Although some researches and management practices have been carried for EFs, most of them are not based on international general concept of EFs. In most cases, ready-made methods in overseas studies are adopted directly or moderately altered to determine the EFs(Yang et al. 2005). The EFs for a majority of rivers are mainly concentrated in the minimum flow in dry seasons or base flow of rivers, which means, major considerations are given to the demand for human production and living water at downstream reaches, while fewer considerations are given to the EFs required by river ecosystems and whole life process of aquatic organisms(Hou et al. 2007). It is not scientific to simply use methods adopted in overseas studies as there's no unified authority standards and methods for EFs. As there are many rivers in China, with great difference in the natural environment and ecosystems, especially due to different levels of human development and utilization, it is difficult to determine the EFs with uniform standards. Different measures should be taken based on the difference in rivers and ecological protection goals. As for big rivers such as Yangtze River and Yellow River, considerations of EFs should not only be given to the differences among upper reaches, middle reaches and lower reaches, but also should be given to the continuity between upper reaches and lower reaches, the continuity between branches and main stream, as well as the scheduling capabilities of water conservancy project. Therefore, it is necessary to research the EFs methods suitable for different types of rivers and conduct the research on EFs based on the actual situations in China, with reference to research results in developed countries and in combination with the characteristics of rivers in China.

EXPLORATION ON EFS PROBLEMS IN DIFFERENT REGIONS OF CHINA

China is a country with vast territory and a large number of rivers, which show great difference in topography, hydrological characteristics, ecological environment, and the influence of human activities(Fu et al. 2006). The EFs problems of rivers in China are very complex. Different rivers show different ecological and environmental problems due to different influences of natural factors such as climate and human activities. River protection goals should be the first consideration to determine EFs, from two aspects - nature and human society. Therefore, the effective way to determine EFs suitable for rivers in China is to make river zoning and classification based on the zonal distribution of rivers in China. Different methods of EFs are then adopted for different river types and protection goals. The determination of EFs is not only protection for river ecosystem, but also a kind of social choice. The zoning of EFs must give appropriate considerations to river characteristics, socio-economic development, ecological environment and water resources management, especially the management of water conservation, in order to facilitate the research and management practices of EFs.

In this paper, based on the global research achievements and the ecosystems of rivers in China, the EFs are classified into three levels, i.e., whole country, river basin and water function area. The first level is the macro-level of the whole country. Main considerations are the geographical locations of rivers. The whole country is divided into three areas according to climate distribution, underlying surface conditions and the river basin boundaries of the seven major river systems(Smalley 2006). Each area may be further subdivided into river basins. As for geographical position, the water resources, population and economic distribution of China have the following characteristics: the spatial distribution of water resources in China is mainly characterized by excessive water resources in south and short water resources in north; the population is mainly concentrated in the eastern and central regions.The population density in the western region is small, especially in northwestern region; with uneven economic development, the eastern coastal region is developed while the western region is lagging behind, and the ecological environment is fragile in the northwestern region(Liefner and Wei 2011, Selya 1984). Based on the above, the zoning of EFs merges the seven major river systems of China into three significantly different types of region for the determination and analysis of water resources allocation and EFs methods, including southern zone, northern zone and continental zone.

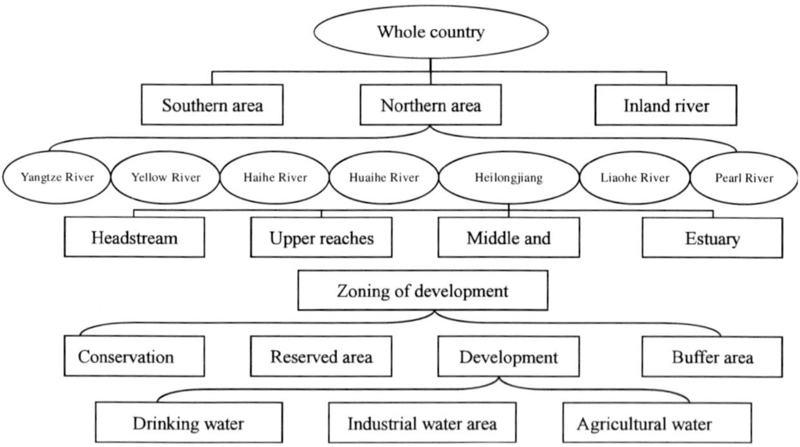

Figure 1. Layered structure of EFs problems in different regions.

The second level is river basin. Main considerations are given to the reaches in the river basin. Different reaches are divided based on the position where the river or reach in the river basin, such as headstream, upper reaches, middle reaches, lower reaches and estuary. There are different aquatic ecological environment problems and management goals at different reaches due to different terrain features and river regimes in their locations; each reach may be further subdivided into different open function areas. The third level is water function area. It is the micro level of protection and management of water resources, with emphasis on the protection of water quality. Different water function areas have different water quality management goals. Water function areas may be further divided into level-one and level-two regions. The level-one region is mainly for solving the water conflicts between different areas, while the level-two region is mainly for solving the water conflicts between different departments. The layered structure of zoning of EFs problems is shown in Figure 1.

Determination Methods of EFs

1. Rivers in the Northern China and Rivers in the Southern China

In the northern China, the reserved water flow is small and may even dry up due to small inflow and large water consumption. Therefore, the river protection goals of northern China are to guarantee the basic or continuous flow for keeping basic river functions(Zhang Y. et al. 2006). Hydrology methods can be used for its calculation, such as Tennant method, Q90 method, and generally 10% is taken as its lower limit (Richter et al. 2006). In some northern areas of extreme water shortage, many rivers have dried up due to large water consumption and there's no way to transfer water from outside basin, the standards may be lowered under the present situations (Song et al. 2007, Zhang Y. et al. 2006). But a long-term planning goal should be set to improve standards and restore the basic functions of the rivers. Water demand for sediment transport should be considered for Yellow River and other sediment rivers(Yang et al. 2009). In northwest China, because the downstream of the inland rivers is often injected into an inland lake, its river protection goals may be to ensure continuous flow; no reduced natural islands areas and sufficient water demand of downstream lakes(Zhang Y. et al. 2006). In the southern China, the EFs may be calculated by hydrology and hydraulics methods. However, because the natural inflow of rivers is large, the standards should be improved accordingly.

Many rivers in the northern China can only maintain basic ecological environment functions or have even lost them(Zhang Y. et al. 2006). On the contrary, the ecological environment is good in southern China due to relatively rich water resources and small proportion of water consumption outside the river in most areas(Zhang N. et al. 2010). Even if the ecological environment in some areas is damaged, it can be restored to good or relatively good state as long as the appropriate protection measures are taken as the total amount of water is large. Therefore, the environmental protection goals of rivers in southern China should be relatively higher than that of in northern China, and its EFs standards should be improved accordingly. Hydrological-biological analysis method and habitat simulation method (such as IFIM method) should be used as far as possible in areas permitted, and even holistic method may be used for calculation in order to meet the water demand and keep the health of ecosystems in rivers(Tharme 2003).

2. Small and Large Rivers

As for some small rivers with deficient data, Tennant method of easy calculation process may be used due to the constraints and less significance of rivers (Tennant 1976). However, the calculation criteria for Tenant method should be determined according to the geographical location of the river. In principle, it is higher in the south and lower in the north. In addition, as for some small rivers, especially some mountainous rivers, because of the great changes in the water inflow season and small water inflow in the dry season, it is difficult to meet the conditions even using the minimum standards of 10% of long-time average annual magnitude by Tennant method. In this case, Q90 method, a method considers the hydrological characteristics of the target with simple calculation process, can be used to determine the EFs in the dry season. But at other times, the flow standards should be improved to meet the annual EFs calculated by Tennant method.

As for large rivers, it is recommended to determine the EFs by holistic method, hydrological-biological analysis method or habitat simulation method (King et al. 2003, Milhous et al. 1984, Tharme 2003). Several factors contribute to this: 1. Large Rivers usually have wide influence and need to be considered from many aspects. 2. Large rivers are of macro-level, providing relatively rich data sources. 3. More manpower, materials and scientific research resources can be mobilized for deeper observation, analysis and research, as many departments attach great importance on large rivers.

3. Polluted River Reaches

As for polluted river reaches, focus should be how much water is needed for self-purification of the river, using methods like Q90 method (improved 7Q10 method), section-beginning control model and water function area method (Cha et al. 2009, Liu et al. 2005, Medina and Water Resources Research Institute of the University of North Carolina. 1982). However, many rivers in China are seriously polluted, they need large amount of water for pollutant dilution, even the whole river may not be sufficient(Tremblay 2006).

Therefore, the management ideas to protect water resources should be changed in actual practice. Water pollution can be controlled by reducing pollution discharge(Shen 2012). To determine the appropriate EFs, several factors should be considered, including the management goals in water function areas, the pollutant carrying capacity of the river (based on the current situation, plans in the short-term and long-term), and pollution discharge control.

4. Special Reaches

Specific conditions should be considered when determining the appropriate EFs for some special reaches. For example, in headstream areas with fragile ecological environment and little impact from human activities, the protection goal is to maintain its natural state as far as possible, taking natural flow as its EFs(Jansky et al. 2011). Estuary areas are often affected by saltwater intrusion, so the amount of water needed to prevent salt water intrusion should be considered(Sun et al. 2012). As for some import reaches, such as rivers with natural reserve of aquatic organisms, simulation calculations (hydrological-biological method, habitat simulation method) should be carried out for EFs, the living characteristics and habitat requirements of aquatic organisms should be also considered. The objective is to ensure the needs of aquatic organisms(Tharme 2003).

Discussions and Conclusion

At present, in China, river conservation mainly focuses on water quality. Measures taken include make water function zoning for rivers, lakes and other water bodies, set specific goals of water quality protection for each water function area. But there's no clear goal for ecosystem protection. There's limited understanding on river ecosystem and few studies in this field (Dudgeon 1995, Ma et al. 2008, Wei et al. 2008). Ecologic protection goals for most rivers are basically in the blank condition, except for some specific reaches, such as rivers in natural reserves where the goals are fish protection, with the aim to protect some particular species or environmental conditions(Tie et al. 2007). Currently, water resource protection is mainly placed on management of water quality and "minimum flow", to ensure water security for people in the downstream (Jiang et al. 2010, Yang et al. 2009). At present, there is a large difference or misunderstanding in the research on EFs in China. A lot of people do not realize that EFs are an entire hydrological process. Even though some people have recognized the importance of EFs process, simple hydrological and hydraulic method is often used when calculating the water needed for protecting ecologic environment and makes external publicity for EFs(Shiau and Wu 2008, Xu et al. 2005). This may result from the limitations imposed by conditions and lack of understanding of the river ecosystem. In the end, the calculation is often a year-round water value rather than the entire flow process, which cannot meet the water demand of river ecosystem in different periods. In addition, habitat simulation method, which using ecological information to establish physical model, is rarely used in China (Yang and Mao 2011, Yang et al. 2009). But it is a dominating method in foreign countries now, and has been used in the practice of the protection and management of water resources(Tharme 2003). Therefore, in China, the focus and development direction of EFs study should be placed on the research of determination methods of EFs, focusing on flow rate and period process in the river for maintaining river health, as well as corresponding physical and chemical characteristics and ecosystem response.

The calculation method of EFs in China must be improved and supplemented. Zoning and classification should be carried out for the river systems and river types in different areas of China. Ecological-physical coupling model should be established for different types of river based on habitat model experiments and long-term river observation data. Different calculation methods should be respectively determined according to the characteristic of rivers. As for important rivers or large rivers, the recommended values of EFs in the river should be determined by holistic method. A team of multidisciplinary experts including water ecologists and water engineers should be set up to get the value in negotiations and discussions of experts based on the results of model experiments and long-term observational data of rivers.

In practice, the situations of rivers are different from each other; the above methods can be used for reference. In some cases there are several factors involved in a river, which lead to different problems to be solved under different monitor data. Compositional methods should be taken for the calculation of EFs. EFs should be determined according to the problems need to be solved. In addition, the determination of EFs is a social choice, so it should consider the development and utilization goals of water resources in the river basin and the region, consider river protection target comprehensively from the point of view of nature and society. The calculation methods of EFs should be determined and calculated based on the protection targets, and carried out the rational allocation of water resources. The choice of determination methods of EFs is not just a purely academic issue, and it should be carried out under the big frame of the allocation and management of water resources. The allocation of water resources is a very complex social process, which may be affected by the interference and substantive participation of parties involved in the process of setting goals, choosing calculation methods and even calculating the EFs of instream flow and outstream withdraw flow, so it is a game-playing process involving various interested parties.

ACKNOWLEGMENTS

This work is funded by CRSRI Open Research Program (Program SN CKWV2012315/KY) and the National Natural Science Foundation of China (Program SN 51109195).

REFERENCES

Cha SM, Ki SJ, Cho KH, Choi H, Kim JH. 2009. Effect of environmental flow management on river water quality: a case study at Yeongsan River, Korea. *Water Science and Technology* 59: 2437-2446.

Dudgeon D. 1995. River Regulation in Southern China - Ecological Implications, Conservation and Environmental-Management. *Regulated Rivers-Research & Management* 11: 35-54.

Fu BJ, Lu YH, Chen LD, Li J. 2006. Progress and prospects of integrated physical geography in China. *Progress in Physical Geography* 30: 659-672.

Hou P, Beeton RJS, Carter RW, Dong XG, Li X. 2007. Response to environmental flows in the Lower Tarim River, Xinjiang, China: An ecological interpretation of water-table dynamics. *Journal of Environmental Management* 83: 383-391.

Jansky B, Engel Z, Kocum J, Sefrna L, Cesak J. 2011. The Amazon River headstream area in the Cordillera Chila, Peru: hydrographic, hydrological and glaciological conditions. *Hydrological Sciences Journal-Journal Des Sciences Hydrologiques* 56: 138-151.

Jiang XH, Arthington A, Liu CM. 2010. Environmental flow requirements of fish in the lower reach of the Yellow River. *Water International* 35: 381-396.

King J, Brown C, Sabet H. 2003. A scenario-based holistic approach to environmental flow assessments for rivers. *River Research and Applications* 19: 619-639.

Liefner I, Wei YD. 2011. Geography and the research on contemporary China Introduction to the Special Issue Foreign direct investment, innovation and regional economic development in China. *Erdkunde* 65: 3-5.

Liu WC, Liu SY, Hsu MH, Kuo AY. 2005. Water quality modeling to determine minimum instream flow for fish survival in tidal rivers. *Journal of Environmental Management* 76: 293-308.

Ma X, Qian J, Xu JC. 2008. Water Resource Management in a Middle Mountain Watershed A Case Study in Xizhuang, Yunnan, China. *Mountain Research and Development* 28: 286-291.

Medina MA, Water Resources Research Institute of the University of North Carolina. 1982. *Hydrologic and water quality modeling for instream flow strategies.* Raleigh, N.C.: Water Resources Research Institute of the University of North Carolina.

Milhous RT, Wegner DL, Waddle T, Western Energy and Land Use Team. Cooperative Instream Flow Service Group., United States. Environmental Protection Agency. 1984. *User's guide to the physical habitat simulation* system (PHABSIM). Washington, DC: The Team.

Richter BD, Warner AT, Meyer JL, Lutz K. 2006. A collaborative and adaptive process for developing environmental flow recommendations. *River Research and Applications* 22: 297-318.

Selya RM. 1984. China - the Geography of Development and Modernization - Pannell,Cw, Ma,Ljc. *Geographical Review* 74: 222-223.

Shen DJ. 2012. Water Quality Management in China. *International Journal of Water Resources Development* 28: 281-297.

Shiau JT, Wu FC. 2008. A histogram matching approach for assessment of flow regime alteration: Application to environmental flow optimization. *River Research and Applications* 24: 914-928.

Smakhtin V. 2007. Environmental flows: a call for hydrology. *Hydrological Processes* 21: 701-703.

Smalley IJ. 2006. Classics in physical geography revisited - Liu Tungsheng 1988: Loess in China (second edition). Beijing: China Ocean Press, Berlin: Springer-Vertag, 224 pp. *Progress in Physical Geography* 30: 673-676.

Song JX, Xu ZX, Liu CM, Li HE. 2007. Ecological and environmental instream flow requirements for the Wei River - the largest tributary of the Yellow River. *Hydrological Processes* 21: 1066-1073.

Sun T, Xu J, Yang ZF. 2012. Objective-Based Method for Environmental Flow Assessment in Estuaries and Its Application to the Yellow River Estuary, China. *Estuaries and Coasts* 35: 892-903.

Tennant D. 1976. Instream flow regimens for fish, wildlife, recreation and related environmental resources. *Fisheries* 1: 6-10.

Tharme RE. 2003. A global perspective on environmental flow assessment: Emerging trends in the development and application of environmental flow methodologies for rivers. *River Research and Applications* 19: 397-441.

Tie W, Mei L, Naiang W. 2007. Protecting China cedar (Cryptomeria fortunei) habitat using GIS-based simulation, modeling of existence probability, and function zoning. *Mountain Research and Development* 27: 352-358.

Tremblay JF. 2006. *China launches huge water quality study.* Chemical & Engineering News 84: 8-8.

Troldborg L, Jensen KH, Engesgaard P, Refsgaard JC, Hinsby K. 2008. Using Environmental Tracers in Modeling Flow in a Complex Shallow Aquifer System. *Journal of Hydrologic Engineering* 13: 1037-1048.

Wei YC, Miao H, Ouyang ZY. 2008. Environmental water requirements and sustainable water resource management in the Haihe River Basin of North China. *International Journal of Sustainable Development and World Ecology* 15: 113-121.

Xu WH, Sandford RC, Worsfold PJ, Carlton A, Hanrahan G. 2005. Flow injection techniques in aquatic environmental analysis: Recent applications and technological advances. *Critical Reviews in Analytical Chemistry* 35: 237-246.

Yang ZF, Mao XF. 2011. Wetland system network analysis for environmental flow allocations in the Baiyangdian Basin, China. *Ecological Modelling* 222: 3785-3794.

Yang ZF, Sun T, Qin XS. 2005. Calculating Methods for Quantifying Environmental Flows in Estuaries: A Case Study of Haihe River Basin, China. *Journal of Environmental Informatics* 6: 72-79.

Yang ZF, Sun T, Cui BS, Chen B, Chen GQ. 2009. Environmental flow requirements for integrated water resources allocation in the Yellow River Basin, China. *Communications in Nonlinear Science and Numerical Simulation* 14: 2469-2481.

Zhang N, Xia ZQ, Jiang H, Huang F. 2010. Influence of the hydraulic engineering construction in the upper reaches on the eco-environmental flow of the middle-lower reaches of the Yangtze River -based on the improved SWAT model. *Technics Technologies Education Management-Ttem* 5: 502-513.

Zhang Y, Yang ZF, Wang XQ. 2006. Methodology to determine regional water demand for instream flow and its application in the Yellow River Basin. *Journal of Environmental Sciences-China* 18: 1031-1039.

In: Yangtze River ISBN: 978-1-62618-285-1
Editors: T. B. Maloney, B. R. Hutchins

Chapter 6

URBAN DEVELOPMENT AND ITS IMPACTS ON ENERGY AND RESOURCE CONSUMPTIONS IN THE YANGTZE RIVER DELTA: TRENDS AND FUTURE PROSPECTS

Keishiro Hara, Michinori Uwasu, Helmut Yabar and Haiyan Zhang

Center for Environmental Innovation Design for Sustainability,
Osaka University,
University of Tsukuba, Japan

ABSTRACT

Due to its rapid industrialization and urbanization, China is facing the challenge of sharply growing energy and resource consumption, while it simultaneously faces various socio-economic challenges such as growing income gaps between urban and rural areas. These challenges are posing threats to regional sustainability. In this paper we look into the Yangtze River Delta region which consists of Shanghai city and the neighboring provinces of Jiangsu and Zhejiang — the fastest economically developing region of the country. We first review recent trends in urbanization and industrialization of the region. We then demonstrate that the rapid urbanization and associated industrial activities have had tremendous impacts on energy and resources consumption, thus having an impact on regional sustainability. We finally discuss that technological innovations in secondary industries and pursuing industrial symbiosis in the context of circular economy would be of vital importance especially for the industrial sectors of the region to pursue sustainable development.

1. INTRODUCTION

China has been experiencing a rapid urbanization and industrialization in the last decades. Urbanization, which is defined in China as the proportion of urban population of the total population, has accelerated significantly in China. In fact, the ratio of people living in urban areas in China increased from 26.44% in 1990 to 35.76% in 2000, and it is predicted to increase to 46.96% in 2010 and even 54.97% in 2020 [UNDESA 2009]. The Yangtze River Delta (Hereinafter YRD), which is located in the eastern part of China, comprises 16 cities in the triangular-shaped area of Shanghai, southern Jiangsu Province and northern Zhejiang Province, making the total area of the YRD about 180,935 km^2. In fact, the YRD is the largest delta among the three distinctive economic deltas including the Pearl River Delta and Beijing-Tianjin-Hebei. The urbanization in the region has been impressive in terms of the scale and speed over the past decades. Such urbanization and associated economic growth would have led to the increase in well-being of people in the region to a certain degree. In the meantime, however, they have also had impacts on the rapid increase in energy and resources consumptions. From the viewpoints of sustainability which is usually argued based upon economic, social and environmental aspects, the recent phenomenon associated with such urbanization and economic growth shall have some implication to regional sustainability.

In this paper, we aim to discuss the implications of such urbanization in the YRD to the expansion of energy and resource consumptions. We first overview the recent trends of urbanization in the YRD region. We then explore the implications that such urbanizations would have in energy and resources (materials) consumptions. Finally we discuss future prospects of the regions particularly from the viewpoints of the industrial activities towards more a sustainable pathway.

2. SCALE OF URBANIZATION IN THE YANGTZE RIVER DELTA

The YRD produces 22.3 % of China's total GDP in 2005 while it only accounts for 2.2 % of China's total land area [Zhang et al 2011]. In fact, the YRD region and two other major delta regions (i.e., Pearl River Delta and Beijing-Tianjin-Hebei) have experienced almost half of the total land use change from arable land to urban built-up areas in China during the period of 1990 and 2000 while the YRD showed the most rapid increase in urbanization among the three delta regions [Zhang et al 2008]. For instance, Shanghai municipality, which is one of the outstanding cities within the YRD, experienced a remarkable increase in the urbanization ratio of 25% during the period of 1990-2000, whereas the ratios for other major cities of Nanjing, Hangzhou and Suzhou are 3.4 %, 5.5 % and 4.1 % respectively. During the period of 2000 and 2005, those cities further experienced a rapid urbanization. For example, the urbanization rate of Nanjing increased by 33.0%, that of Hangzhou increased by 33.2%, and that of Suzhou increased by 17.9%. [Zhang et al 2011]. It is observed that the rate of urbanization in the YRD was higher for coastal cities than for inland cities. As the Chinese government now explores further economic development even in the inner areas, there is a possibility that cities located in the inner areas will be economically developed and urbanized.

Importantly, urbanization and land use change are clearly associated with socio-economic development. Specifically, the YRD core cities, such as Shanghai, Nanjing, Hangzhou and Suzhou, experienced a relatively high increase in socio-economic conditions. These core cities attained not only high socio-economic development along with moderate environmental pressure. Indeed, such urbanization in general tends to bring about various challenges, as well. Such challenges include, but are not limited to pollutions, overconsumption of energy and resources and societal inequality between the urban and rural areas. Above all, overconsumption of energy and resources driven by urbanization has a clear implication to regional sustainability. In the following section, we address the aspect of increasing consumption of energy and resources by looking into the case of YRD.

3. Growing Energy and Resource Consumptions – Recent Trends

As the urbanization has been taking place at an unprecedented pace, demands for resources and materials indispensable for developing urban infrastructures have been growing. Indeed, productions of cements and steels, the essential materials for setting infrastructures, have been impressively increasing in China. An empirical study clearly indicates, both cement and steel production increased exponentially in relation to urbanization ratio in China [Shen et al. 2005], implying that urbanization is accompanied by the massive increase in the consumptions of such materials. In fact, crude steel production in the Jiangsu Province between 1997 and 2007 grew by almost ten times, from 4.75 to 47.21 million tons, while cement production during the same period expanded from 40.31 to 118.50 million tons [NBSC 2008]. These materials produced have been stocked in the urban infrastructures in the various forms. Han and Xiang (2012) have estimated ten types of materials stocked in four major infrastructures such as residential building, road, railway, and water pipeline in 31 provinces during 1978–2008, and concluded that the total volume of materials stocked in infrastructures has increased at a speed of 8.5 % per annum and reached 42.5 billion tons by the end of 2008 while its per capita value increased by nine times compared with that of 1978. This indicates that massive amount of resources have been used to produce various materials which are accumulated in the urban areas. The stocked materials in urban areas shall continue to grow in coming years along with economic growth and urbanization.

With the driving force of urbanization, the YRD indeed accounted for 22% of the country's gross secondary industrial production outputs [Chinese Statistical Bureau 2005]. While such secondary industries play a crucial role in the economic activities in China, cement and steel making industries represent the most energy-intensive industries in China, contributing to the overall energy consumption in the country. This also means that growing cement and steel productions will inevitably lead to a rapid increase in energy consumption in China. Furthermore, the increase in energy consumption will eventually lead to the growth of CO_2 emissions. Given that the cement and steel productions shall continue to grow driven by the urbanization and economic activities, improving energy efficiency in the production processes will be a major challenge in the years to come. Between 1980 and 2000, China quadrupled its GDP while its energy consumption only doubled, achieving the energy intensity improvement by 50 % [NBSC 2008]. In particular, the marked improvement in energy intensity was achieved in energy-intensive industries including the cement production [Liao et al 2007]. However, in its 10th five-year plan between 2001 and 2005, the energy intensity started to increase after 2001. Studies explained that such reversed trend of energy intensity is attributed to two main factors: i.e., 1) significant structural changes within the industrial sector [Ma and Stern 2008] and 2) a decline in the ratio of industrial technological innovations [Andrew-Speed 2009]. With regards to the structural change, expansion of secondary industry, which had been driven by the urbanization and economic growth, has focused on energy intensive sector, such as cement and steel making industries. Notably, energy efficiency gains achieved by technological developments were rather small in comparison with the increases in the productions within those energy-intensive sectors. Hara et al (2011) studied energy intensity and consumption trends in Shanghai city and Jiangsu province and concluded that both GDP and energy consumption levels have been growing at a much higher pace than the possible pathways towards the government's mid-term targets of quadrupling GDP while doubling energy consumption by 2020, under the Business as Usual (BAU) development patterns both in Shanghai city and Jiangsu province.

Importantly, it is found that there is a room for an improvement in technologies used in the production processes of energy-intensive industries, such as cement and steel making factories. There are still many small-size factories with outdated technologies and this situation usually results in the poor performances of energy efficiency. Hara et al (2011) calculated the potential energy savings associated with steel making industries in Jiangsu province and concluded that the potential energy savings by replacing the current technologies with the best available technologies are equivalent to almost 3.25 million tce (tons carbon equivalent), which is estimated to be almost 15% of the total energy consumption in Jiangsu province as of 2008. This fact demonstrates that the technological improvement in industrial sectors is of great help in reducing energy consumption and increasing in energy efficiency in the industrial sectors in China.

CONCLUSION - PROSPECTS FOR THE FUTURE

As urbanization is likely to continue in the YRD region, the impacts of urbanization on regional sustainability shall be continuously growing. This paper particularly discussed the linkages of such urbanization with growing energy and resource consumption. As the total consumption might grow anyway in accordance with continuing urbanization, it is essential to pursue the improvement in efficiency of energy and resource use. Though the Chinese government sets the mid-term policy target to achieve 50 % improvement of energy intensity by 2020, it appears that achieving the goal is not an easy task. While various measures are needed to achieve energy and resource efficiency, it would be highly essential to change the development course of secondary industry which still constitutes the core economic activity in China.

Hara et al (2011) argue that it important for China to design strong and effective policies that will facilitate the replacement of outdated production technologies with new and highly energy-efficient technologies. It is also essential that the energy-intensive secondary industries attempt to facilitate the practices of industrial symbiosis and promote resource-circulation system which would help reduce the amount of raw materials consumption. For example, plastic wastes from urban areas can be utilized in the steel industry. Another example is the utilization of sewage sludge in cement industry. In Japan, inclusions of sewage sludge as a raw material increased from 2.65 million tons in 2004 to 3.04 million tons in 2008 [Japan Cement Association 2006]. Industries are encouraged to advance both by-products exchange among industries and the use of wastes from the urban sector, creating the loop-closing systems and a circular economy. China actually introduced the concept of a ‘‘circular economy’’ (CE) as “The Chinese Circular Economy Promotion Law’’ became active in January 2009. The objective of the CE model is to promote resource circulation at the industry level (small-sized circulation), the eco-industry level (mid-sized circulation) and the city/provincial level (regional circulation) to maximize the potential of resource utilization [Hara et al 2011]. Promoting industrial symbiosis and loop-closing systems linking the industry and urban systems would be the key in line with the governmental policy.

Beside such measures at industry level, promoting integrative approach is of critical importance. Among others, incentivizing the public to change their life styles shall be indispensable so that sustainable consumption is facilitated. In this regards, sustainable production and consumption shall be the key to pursue sustainable development not only in the YRD region, but also in China as a whole.

REFERENCES

Andrew-Speed P (2009). China's ongoing energy efficiency drive: origins, progress and prospects. *Energy Policy,* Vol. 37, 1331–1344.

Chinese Statistical Bureau (2005). China statistical year book for regional economy. China Statistics Press, Beijing.

Han J & Xiang W.J (2012). Analysis of material stock accumulation in China's infrastructure and its regional disparity, *Sustainability Science*, in press.

Hara, K., Yabar, H., Uwasu, M & Zhang, H (2011). Energy Intensity Trends and Scenarios for China's Industrial Sectors - a Regional Case Study, *Sustainability Science*, Vo. 6 (2), 123-134.

Japan Cement Association (2006) Trends of by-products and wastes utilization in cement industry, (Japan Cement Association Home Page: *http:// www.jcassoc.or.jp/cement/2eng/ea.html*, accessed on Jan 31, 2010) (in Japanese)

Liao, H., Fan, L & Wei, Y.M (2007). What induced China's energy intensity to fluctuate: 1997–2006? *Energy Policy,* Vol.35, 4640–4649.

Ma C, Stern D (2008). China's changing energy intensity trend: a decomposition analysis. *Energy Econ*, Vol. 30, 1037–1053.

National Bureau of Statistics of China (2008). China statistical yearbook 2009. China Statistics Press, Beijing.

National Bureau of Statistics of China (2008). China statistical yearbook 2009, China Statistics Press, Beijing.

United Nations, Department of Economics and Social Affairs (2009). World Urbanization Prospects: the 2009 review. UNDESA. POP/DB/ WUP/Rev.2009/1/F2.

Zhang, H., Uwasu, M., Hara, K & Yabar, H (2011). Sustainable Urban Development and Land Use Change—A Case Study of the Yangtze River Delta in China. *Sustainability. Vol. 3(7),* 1074-1089.

Zhang, H., Uwasu, M., Hara, K., Yabar, H., Yamaguchi, Y & Murayama, T (2008). Analysis of land use changes and environmental loads during urbanization in China. *J. Asian Archit. Build. Eng. 7*, 109-115.

INDEX

A

access, 51, 52, 78
accessibility, 46
accounting, 100
acid, 47
additives, 4, 24
adverse effects, 10, 20, 27
Afghanistan, 43, 60
agencies, 71
agriculture, 2, 50, 101
air temperature, 36, 39, 43, 49, 54, 63
Alaska, 30
amplitude, 49
ancestors, 56
aquatic life, 56
aquatic systems, 30
aromatic hydrocarbons, 33
arsenic, 51
arteries, 2
Asia, vii, viii, 36, 63, 65, 67, 87, 90
assessment, viii, ix, x, 2, 20, 27, 28, 60, 63, 66, 68, 71, 72, 76, 78, 84, 86, 88, 118, 119
atmosphere, 3, 4, 30
atmospheric deposition, 32
atoms, 23
authority, 110
avoidance, 78

B

bankers, 83
banks, 50, 90, 91
base, ix, 47, 66, 86, 110
basic education, 78
Beijing, 29, 32, 61, 64, 87, 93, 118, 122, 123, 127
benefits, 68, 69, 76, 85
bias, 71, 72, 76, 83
bioaccumulation, vii, 1, 27, 28
biodegradability, 15
biodiversity, 57
biological processes, 23, 28
biomass, 54, 64
by-products, 126, 127

C

cadmium, 51
canals, 90
carbon, 53, 84, 125
carbon dioxide, 84
case study, 60, 117
Central Asia, 43
challenges, xi, 69, 121, 123
chemical(s), 10, 27, 28, 29, 31, 100, 110, 116
chemical characteristics, 116

Chinese government, 123, 126
chlorination, 27
chlorine, 25, 26
circulation, 41, 126
city(ies), viii, 2, 35, 36, 38, 65, 67, 70, 78, 90, 91, 95, 122, 123
civil war, 92
civilization, 37, 106
classification, x, 109, 111, 116
climate, viii, 35, 36, 38, 39, 41, 47, 48, 49, 52, 54, 59, 60, 62, 77, 84, 90, 100, 111
climate change, viii, 35, 52, 54, 59, 60, 77, 84, 100
closure, 69
CO_2, 124
coal, 77
coastal region, 112
cobalt, 55
commercial, 2, 8, 18, 25
community(ies), 64, 71, 78, 82, 85
compensation, 56
compilation, 4
complexity, 23, 28, 68, 79, 85
composition, 9, 15, 20, 24, 25, 28, 53
compounds, 26, 28
conflict, 85
connectivity, viii, 65, 67
consensus, 39
conservation, 67, 111, 115
conserving, 59
construction, viii, 35, 36, 56, 68, 69, 76, 77, 78, 84, 92, 94, 95, 106, 120
consumption, xi, 2, 29, 113, 121, 123, 124, 125, 126
contaminant, 3, 10, 97, 98, 102
contamination, vii, 1, 2, 3, 8, 10, 12, 15, 23, 24, 27, 29, 32, 63
cooling, 39, 55, 64
copper, 55
cost, 105, 106
covering, 77, 100
criticism, 69, 85
crop(s), 53, 78
culture, 87

D

danger, 88, 97
data set, 39
DDT, viii, 2, 3, 4, 5, 6, 7, 8, 9, 10, 12, 14, 15, 18, 22, 28, 30, 31, 32, 33
decision makers, 70
decomposition, 10, 103, 127
deer, 46
deficit, 94
deflation, 38, 50
deforestation, 37, 55
deformation, 62
degradation, 3, 4, 10, 15, 22, 29, 30, 48, 61, 62, 63, 103
delta, vi, vii, xi, 33, 90, 121, 122, 123, 127
deposition, 3, 4
deposits, 50, 51, 104, 105
depth, 38, 41, 49, 64, 90
destruction, 43, 58, 85
detection, 14
detention, 93, 96, 102
developed countries, 111
developing countries, 87
dibenzo-p-dioxins, 32
diffusion, 26
direct investment, 118
disaster, 100, 103, 104, 105, 106, 109
discharge volume, viii, 65, 67
discharges, 60, 98, 103
displacement, 56, 59, 92
dissolved oxygen, 25
distribution, viii, 2, 4, 8, 14, 24, 28, 30, 32, 33, 43, 48, 49, 111
DOI, 60
drainage, 39, 52, 57, 92
drinking water, ix, 66, 70
drought, 97, 101, 102, 103, 106
dry matter, 53
drying, 57, 62

E

earthquakes, 50, 53, 57

East Asia, 37, 64
East China Sea (ECS), vii, 1, 2, 9, 14, 15
ecological information, 116
ecology, 57
economic activity, 126
economic development, 68, 88, 111, 112, 118, 123
economic growth, 122, 124, 125
ecosystem, vii, ix, x, 2, 3, 4, 10, 27, 28, 61, 66, 70, 82, 83, 84, 86, 92, 93, 103, 109, 110, 111, 115
ecotoxicological, viii, 2, 27, 28
electricity, viii, 36, 56, 59, 68, 69
emission, 10, 29
employment, 78
enantiomers, 4, 22, 23, 28
endangered, 46, 84
endangered species, 52, 77
energy, vii, xi, 55, 56, 92, 121, 122, 123, 124, 125, 126, 127
energy consumption, 124, 125
energy efficiency, 124, 125, 127
energy supply, 56
enforcement, 71
engineering, 62, 67, 120
environment(s), vii, x, 1, 2, 3, 4, 9, 19, 20, 22, 23, 25, 26, 28, 29, 36, 37, 53, 62, 63, 64, 70, 84, 109, 110, 111, 112, 113, 115, 116
environmental aspects, 122
environmental change, 36
environmental conditions, 2, 10, 20, 115
environmental impact, 61, 63, 68, 88
environmental implications, vii, x, 63, 66, 86
environmental issues, 85
environmental problems, x, 109, 111
environmental protection, 68, 69, 71, 113
Environmental Protection Agency (EPA), 27, 118
environmental quality, 3
environmental resources, 119
environmental stresses, 100
environmental sustainability, x, 89
equilibrium, 48
equity, 70, 71
erosion, viii, 36, 47, 50, 54, 56, 61, 77, 84
evaporation, 41, 47, 57
evapotranspiration, 41, 60
evidence, 41, 48, 49
evolution, 62, 64, 110

F

factories, 125
families, 43, 78
farmers, 51
farmland, 50, 78
farms, 54
feelings, 68
fencing, 54
fish, 10, 25, 115, 118, 119
fission, 63
flammability, 4
flocculation, 8
flooding, 56, 69, 78, 85, 91, 100, 110
floods, x, 78, 86, 89, 90, 91, 92, 93, 95, 96, 100, 103, 104, 105
flora, 43
fluid, 4
folklore, 78
food, vii, 1, 3, 10, 12, 31, 46
food web, vii, 1, 3, 10, 12, 31
forbs, 43, 54
force, 72, 124
forest resources, 52
formation, 61
freezing, 49, 61
freshwater, 90, 91, 94, 95, 101
frost, 39, 41, 47, 49

G

GDP, ix, 56, 66, 69, 90, 123, 124
GEF, 32
geography, vii, 117, 118
geology, 84
Germany, 63
GIS, 119

governments, ix, 65, 66, 70, 86
GPS, 64
grades, 47, 50
grading, 43
grass, 53
grasslands, 52, 54, 58
gravity, 99
grazing, 37, 43, 52, 53, 54, 55, 59, 62, 64
Great Leap Forward, 55
greenhouse gas emissions, 77
gross domestic product, 56
groundwater, 49, 60
growth, 54, 59, 100, 103, 106, 122, 124
guidance, xi, 69, 110
guidelines, 10, 27, 28

H

habitat, ix, 53, 66, 70, 110, 113, 114, 115, 116, 118, 119
happiness, 78, 85
harmony, 53
harvesting, 58
hazards, 53
health, vii, 2, 78, 114, 116
health care, 78
hexachlorocyclohexanes, viii, 2, 32
highways, 36
histogram, 118
history, 76, 107
horses, 53
housing, 52
human, vii, x, 1, 3, 4, 10, 27, 28, 98, 99, 100, 104, 109, 110, 111, 115
human activity, 104
human development, x, 109, 110
human health, 3, 4, 10, 27, 28
humidity, 41
hunting, 52
hydrocarbons, 29, 31
hydroelectric power, 37, 56

I

ideal, 4, 70, 85
identification, 78
identity, 83
images, 77
immigration, 78
impact assessment, 70, 87, 88
improved riparian management, ix, 66, 70
improvements, ix, 66, 69, 76
income, xi, 121
industrial sectors, xi, 122, 125
industrialisation, 78, 85
industrialization, xi, 121, 122
industry(ies), xi, 57, 101, 122, 124, 125, 126, 127
inequality, 123
infrastructure, 78, 127
insecticide, 9
integrity, 71, 83
interdependence, 71
interference, 117
intrinsic value, ix, 66
investment, 106
iron, 55
irrigation, 67, 69, 78, 87, 100
islands, 113
isolation, 70
isomerization, 19
isomers, 9, 18, 20, 29, 30, 31, 32
issues, ix, 66, 69, 70, 78, 85, 86, 100
Italy, 29

J

Japan, 30, 33, 121, 126, 127
justification, 86

K

Korea, 30, 117

L

lakes, ix, x, 38, 47, 50, 66, 70, 89, 91, 93, 96-102, 104, 107, 113, 115
landfills, 26
landscape, 43, 46, 47, 50, 56
Late Pleistocene, 60
leaching, 15
lead, 3, 28, 51, 55, 116, 124
levees, 91, 97, 99, 102, 106
lifetime, 28
livestock, 54
local government, x, 89, 95
logging, 52
lying, 47

M

magnesium, 56
magnitude, 114
MAI, 88
majority, 99, 110
mammals, 46
management, ix, 32, 66, 67, 70, 86, 96, 107, 110, 111, 112, 115, 117
manganese, 55
manpower, 114
marine environment, 3
mass, 29, 47, 49
materials, 4, 114, 122, 124
matrixes, 3
matter, 3, 8, 23, 53
measurement(s), 64, 70, 76, 85
media, 22, 23
median, 10, 14, 16, 17, 18, 21, 22
Mediterranean, 43
melt, 54
melting, 49
mental model, 71
mercury, 56
metabolites, 9, 10, 28
metaphor, ix, 66, 70, 86
methodology, vii
migration, 20, 63
military, 43
minorities, 51
minority groups, 59
misunderstanding, 115
mixing, 99
models, 107
moisture, 44, 49
Mongolia, 43, 64
mortality, 54
motivation, 85
mountain ranges, 38
multivariate analysis, 25, 28
museums, 77

N

National Academy of Sciences, 60
natural resources, 100
negative effects, 54
neutral, 47
New Zealand, ix, 65, 66, 71, 72, 85, 86, 87
nitrification, 95
nitrogen, 53
NOAA, 27, 31
North America, 43
nutrient, 53, 98, 100

O

oil, 36
one dimension, 72
operations, 55, 56
opportunities, x, 67, 78, 82, 85, 87
optimization, 118
ores, 55
organ, 72
organic matter, 10, 23, 25, 27, 47
organochlorine compounds, 32
organochlorine pesticides (OCPs), vii, 1
overgrazing, 44

P

Pacific, x, 30, 38, 63, 67, 87

parallel, 52, 72
partition, 3
pasture, 63
pathways, viii, 2, 3, 125
payback period, x, 89, 106
PCA, 25
PCBs, viii, 2, 4, 21, 23, 24, 26-28, 30, 32
peace, 61
Pearl River Delta, 33, 122, 123
permafrost, 36, 39, 47, 48, 49, 50, 60-64
persistent organic pollutants (POPs), vii, 2, 3, 27, 32
personal communication, 39
pesticide, 2, 30
pH, 47
photographs, 59
physicochemical characteristics, 27
physicochemical properties, 4
pipeline, 36, 124
plants, 36, 43, 48, 77, 84
poison, 54
policy, 54, 126
pollutants, vii, 2, 3, 8, 10, 28, 31, 32, 101, 107
pollution, vii, viii, x, 2, 3, 8, 14, 24, 26, 27, 28, 31, 51, 69, 71, 77, 89, 91, 93, 94, 96, 98, 99, 101, 104, 106, 107, 115
polychlorinated biphenyl(s) (PCBs), vii, 1, 2, 3, 31, 33, 34
polycyclic aromatic hydrocarbon, 33
poor performance, 125
population, viii, 37, 51, 54, 57, 59, 65, 67, 90, 100, 106, 111, 122
population density, 59, 112
poverty, 54
power generation, ix, 66, 68, 70, 77, 86
Poyang Lakes, x, 89
precipitation, 26, 36, 37, 39, 41, 42, 47, 50, 90, 91, 96, 99, 100, 102
prevention, x, 89, 102, 107
principal component analysis, 25
probability, 119
producers, 20
project, 51, 57, 67-70, 72, 76, 77, 78, 82, 83, 84, 85, 86, 93, 94, 95, 104, 111
protected areas, 52
protection, x, 37, 68, 89, 90, 99, 102, 107, 109, 110, 111, 112, 113, 115, 116
public health, 78
pumps, 26
purification, 103, 114

Q

Qinghai-Tibetan Plateau, viii, 35

R

race, 4, 22, 28
radiation, 41
radius, 77
rainfall, 91
rangeland, 53, 62
rate of change, 39
raw materials, 26, 126
reality, 75, 92
recognition, 34
recommendations, 118
recreation, 119
recurrence, 78
recycling, 2, 32
redistribution, 20
reform(s), 54, 59
regrowth, 59
relevance, ix, 66, 70, 71, 72, 77, 86
reliability, ix, 66, 70, 86
relief, 102
religion, 54
reporters, 60
requirements, 72, 115, 118, 119
reserves, 56, 115
resettlement, viii, 36, 51, 56
residues, 4, 8, 19, 20, 28, 30
resource management, ix, 66, 86, 119
resource utilization, 126
resources, xi, 82, 92, 111, 114, 117, 121, 122, 123, 124
response, 63, 116

risk(s), ix, 3, 4, 10, 12, 26, 28, 60, 66, 69, 78, 95
risk assessment, 3, 4, 26, 28
river basins, 111
river flows, ix, 38, 66, 70, 86, 91
river systems, 111, 116
routes, 57
runoff, 2, 3, 26, 59, 90, 94, 103, 104
rural areas, xi, 121, 123

S

safety, 78
saltwater, 115
savings, 125
science, 87
sea level, 90, 93, 101, 104
security, 53, 70, 78, 115
security guard, 53
sediment, ix, 2, 3, 4, 8, 9, 10, 20, 24, 25, 26, 27, 28, 29, 31, 33, 51, 56, 61, 66, 67, 69, 96, 98, 101, 102, 103, 104, 105, 113
sedimentation, x, 8, 69, 77, 89, 92, 96, 97, 104, 105, 106, 107
sediments, vii, 2-12, 14, 15, 16, 18-31, 33, 34, 51
seedlings, 59
sensitivity, 72, 76, 83, 84, 85
settlements, 43, 51, 57
sewage, 8, 14, 23, 26, 28, 29, 126
sheep, 53
shoreline, 97
shortage, x, 37, 89, 93, 97, 103, 104, 108, 113
showing, 40, 49, 60
shrubland, 54
shrubs, 53
signs, 77
simulation, 110, 113-116, 118, 119
sludge, 26, 29, 126
society, 78, 82, 111, 116
soil erosion, 59
solubility, 3, 15
solution, 101, 104
sorption, 8
South Asia, 37
Southeast Asia, 37
species, ix, 43, 46, 53, 66, 70, 84, 115
spring, 49, 64
stability, ix, 4, 64, 66, 69, 84
stakeholder groups, 74
stakeholders, 71
standard of living, ix, 66, 69
state, 54, 76, 82, 113, 115
statistics, 23
steel, 124, 125, 126
steel industry, 126
stock, 53, 54, 58, 127
storage, 69, 92, 97, 98, 100, 103-105, 107
structural changes, 125
structure, 62, 112, 113
structuring, 75
subjectivity, 75
supervision, 52
surface area, 101
survival, 46, 78, 118
sustainability, vii, ix, x, xi, 66, 69-72, 74, 75, 76, 82, 84, 86, 105, 121-123, 125
sustainable development, xi, 68, 119, 122, 126
symbiosis, xi, 122, 126

T

Taiwan, 30
target, 114, 116, 126
techniques, 68, 119
technological advances, 119
technological developments, 125
technology(ies), x, 67, 89, 125, 126
temperature, 39, 41, 49, 61, 64
tension(s), ix, 66, 69, 85, 87
territory, 95, 111
threats, xi, 92, 121
Tibet, 36, 37, 43, 49, 51, 60-64, 90
Tibetans, 51, 53, 54
time frame, 86
time periods, 70
time series, 41
tin, 55

titanium, 56
total energy, 125
tourism, 77, 100
toxicity, 11, 29, 32
trade, ix, 66, 69, 78
traditions, 84
transmission, 26
transparency, 71
transport, 4, 113
transportation, vii, viii, ix, 1, 3, 4, 47, 52, 65, 66, 67, 70, 86
trial, 68
tungsten, 55
turbulence, 103

U

UNDP, 68
UNESCO, 52, 53
uniform, x, 109, 110
United Nations, 127
United States (USA), 30, 87, 88, 118
urban, ix, xi, 24, 30, 66, 70, 86, 121, 123
urban areas, vii, 1, 101, 122, 124, 126
urban population, 122
urbanisation, 78
urbanization, xi, 100, 121, 122. 123. 124,125, 127

V

vacuum, 26
vanadium, 56
vapor, 15
variables, 25
vegetation, viii, 35, 37, 42, 43, 47, 50, 53, 55, 58, 59, 60
velocity, 98
vulnerability, 63

W

waste, 2, 4, 29, 30, 32
wastewater, 3, 28, 95, 96, 98, 99, 107
water quality, 51, 77, 92, 98, 100, 101, 103, 112, 115, 117, 118, 119
water resources, 100, 106, 111, 112, 113, 115, 116, 119
water shortages, 100
watershed, 32, 58, 61
well-being, ix, 66, 71, 72, 122
wetlands, ix, 66, 70, 101
wetting, 62
wildlife, 77, 84, 119
wind speed, 41
World Bank, 68
worldview, 76, 82, 83
worldwide, 4

Y

Yangtze River Estuary (YRE), vii, 1, 2
yield, 39, 61, 64
yin, 72
yin-yang, 72
yuan, 105, 106

Z

zinc, 51, 55